# From Bunker Hill to the Boardroom: Battle-Tested Leadership for Today's Executives

Andy Moye, Ph.D.

First Hardback and Paperback Edition



Published by Freiling Agency, LLC.

P.O. Box 1264
Warrenton, VA 20188

www.FreilingAgency.com

HB ISBN: 978-1-969826-44-3
PB ISBN: 978-1-969826-43-6
E-book ISBN: 978-1-969826-45-0

# *Dedication*

*For my family*

# Contents

# *Foreword*

## *By General Stanley McChrystal (Ret.)*

WHEN PEOPLE ASK WHETHER leadership lessons from war apply beyond the battlefield, I often find myself thinking about Abraham Lincoln and Ulysses S. Grant.

In March of 1864, Lincoln made a decision that many in Washington found unsettling. He promoted Grant to lieutenant general and placed him in command of all Union armies, the first officer to hold that rank since George Washington. Grant was not polished, nor was he politically adept. He had failed in business, was rumored to drink too much, and did not present himself as the kind of general who reassured capital cities. But he moved. And after three years of war defined by hesitation and retreat, movement mattered.

What Lincoln did next was more consequential than the promotion itself. He largely left Grant alone.

That restraint did not come easily. The war was unpopular, casualties were mounting, and political pressure in Washington was relentless. Grant's Overland Campaign produced staggering losses. Newspapers called him reckless. Members of Congress urged Lincoln to replace him. Yet Lincoln resisted the urge to manage the war from his desk. He understood that the problem was not the absence of oversight but the absence of alignment. Grant understood the nature of the conflict and was willing to bear its cost. Lincoln chose to bear his own cost in the political arena and allow Grant to fight.

After the Battle of the Wilderness, Lincoln wrote that Grant was the first general he could trust to do the right thing. It was not praise for a single maneuver. It was recognition of judgment.

The two men did not need constant meetings or detailed directives. They shared a common understanding of the objective and of the burden required to reach it. The war ended not because either man controlled every variable, but because authority and responsibility were properly placed and then honored.

That relationship has stayed with me.

In Iraq and Afghanistan, we operated in environments that shifted faster than any headquarters could fully grasp. Our early instinct was to increase control, to move decisions upward where information seemed more complete. That approach made us slower and, in some cases, less effective. Over time we learned that clarity of intent mattered more than proximity to authority. When people understood the purpose of their actions and were trusted to exercise judgment, the organization moved with greater coherence than any centralized model could produce.

When I left the military and began working with businesses and institutions outside of government, I encountered similar patterns. The pressures were different, but the tension was the same. Leaders are asked to manage risk while encouraging initiative. They are expected to maintain standards while adapting to change. The instinct to centralize authority is strong, particularly when outcomes are uncertain. Yet the more complex the environment, the less effective rigid control becomes.

This is why *From Bunker Hill to the Boardroom* resonated with me.

Andy Moye does not treat leadership as a collection of techniques nor a modern discovery. He examines it as a recurring human responsibility shaped by context and consequence. By following it across generations, he illustrates how the outward form of leadership changes while the internal demands remain constant. Leaders must provide clarity about what matters. They must respond when circumstances invalidate their assumptions. They must distribute responsibility without abandoning accountability.

What I found compelling is the seriousness with which the book approaches these obligations. It does not reduce leadership to optimism or charisma. It acknowledges that judgment is imperfect and that decisions carry cost. It recognizes that trust is earned slowly and can be lost quickly. It treats empowerment not as a slogan but as a discipline that requires restraint and patience.

Leadership, in my experience, is less about directing every action than about shaping conditions in which capable people can act well. That idea runs quietly through these pages. The historical episodes are not presented as heroic legends but as moments of choice, often made under pressure and without full information. The parallels to contemporary organizations are clear without being forced.

For those who have carried responsibility for others, whether in uniform or in civilian life, this book will feel grounded in reality. It does not promise certainty. It offers perspective. And perspective is often what leaders lack when circumstances narrow their field of vision.

The burden of leadership has never been light. It is measured not only in results but also in the character required to pursue them. This book reflects that understanding with clarity and restraint. That is why it stayed with me.

# *Introduction*

APRIL 6, 2003. SOMEWHERE over northern Iraq, 25,000 feet above a landscape that had become the latest theater in America's ongoing war on terror. I was an O-3 lieutenant in the U.S. Navy—equivalent to a captain in the other services—serving as tactical coordinator aboard a P-3C Orion aircraft. Our call sign was CAC-4, Combat Air Crew Four, and we were the premier crew in Patrol Squadron Five, the ones they sent first on the really hard missions.

The P-3 isn't what most people picture when they think of military aircraft. It's not a sleek fighter jet or a massive bomber. It's a four-propeller workhorse originally designed to hunt Soviet submarines during the Cold War. But in 2003, with the submarine threat diminished and unmanned drones still in their infancy, the Navy had retrofitted these aircraft for overland Intelligence, Surveillance, and Reconnaissance operations—ISR, in military parlance.

Our crew of eleven had been flying almost daily missions since the shock-and-awe invasion began in late March. We'd take off from Souda Bay in Crete, fly through Turkish airspace, and cross into northern Iraq—territory that had never before seen American ISR aircraft. The missions were grueling: executing twelve-hour flights over hostile territory, constantly watching for surface-to-air missiles, searching for weapons of mass destruction, and hunting high-value targets from Saddam's deck of playing cards.

But it wasn't the external threats that had me most concerned that night. It was my crew.

Despite being handpicked from the top of their respective classes and specialties—the best pilots, the most skilled sensor operators, the most experienced flight engineers—they were rapidly losing faith. The daily grind of combat missions, the stress of flying non-maneuverable aircraft through missile-infested airspace, the mounting fatigue from weeks of continuous operations—it was all taking a toll. As a young leader, I was struggling with a fundamental challenge: How do you keep a team motivated when hope seems lost? How do you maintain purpose when every day feels as if you're just going through the motions?

Two hours into our mission that night, everything changed.

The call came over the radio, something we'd never heard before: "P-3, you're in position to support a search and rescue. An Army helicopter has gone down near Baghdad. Search and rescue forces are en route, but we need an eye in the sky to help locate that downed aircraft and ensure we can rescue those aviators."

I keyed the intercom and relayed the mission change to my crew. The response was immediate and jarring—I nearly had a mutiny on my hands. The answer from almost everyone was a resounding "No!" The closer we'd get to Baghdad, the higher the chance of getting shot down. An aircraft had just been shot down outside Baghdad. Why would we risk eleven lives for what might already be a lost cause?

For about ten minutes, with everyone at their stations and the radio crackling with urgency, I had to do something I'd never done before. I had to remind my crew—my team, my brothers—of their purpose.

"Hey guys," I said over the intercom, "what are we doing here? We're here to support the United States. We all took an oath to the Constitution. We all took an oath to defend this country. We all took an oath to defend each other. Right now, our brothers and sisters are on the ground in Baghdad, and we have an opportunity to support them."

Something shifted in that moment. We could feel it through the aircraft—a renewed sense of purpose, a reconnection to something larger than our individual fears. We spent the next four to five hours supporting search-and-rescue operations outside

Baghdad, providing overwatch, identifying threats, and staying on station until we had to shut down an engine to conserve fuel and finally land at an airfield in Saudi Arabia.

When we touched down and completed the mission debrief, I could see it in everyone's eyes—that renewed sense of purpose. Crew members who had been slowly losing faith in their mission just hours earlier were energized, focused, and proud of what they'd accomplished. We'd helped find that downed aircraft. Everyone survived.

That night taught me three leadership lessons that have guided me ever since:

**First, leaders don't just provide vision—they also provide purpose.** Vision tells people where they're going; purpose tells them why it matters. In that moment over Baghdad, I didn't need to paint a picture of some distant strategic objective. I needed to connect my crew to the immediate human reality: Fellow servicemembers were in danger, and we were the only ones in position to help.

**Second, leaders must cultivate adaptive intelligence.** We had launched that night with a specific mission—hunting weapons caches and gathering intelligence on predetermined targets. But war doesn't respect a flight plan. When circumstances changed, we had to think on our feet, rethink our approach, and execute a mission we'd never trained for in one of the most dangerous airspaces in the world.

**Third, leaders must empower their teams.** I couldn't fly that mission alone. Every person on that aircraft—from the pilots to the sensor operators to the flight engineers—had to own the mission. They all had to feel that if those soldiers died on the ground, it was on them personally. That's the principle of decentralized command: If a bullet had come through the aircraft and killed me, all who were left should have been able to finish the mission and know exactly what they were supposed to do.

Purpose. Adaptive intelligence. Empowered teams.

Twenty years later, those three principles continue to guide my leadership approach, whether I'm running a healthcare technology company, leading a startup through a pivot, or managing teams building AI solutions. But it wasn't until recently that I realized these weren't just lessons from modern warfare—they were part of something much deeper, something that had been passed down through generations of my family.

It started with my son's curiosity about our heritage. A few years ago, when he was sixteen, he became fascinated with where the Moye name came from. We'd never really explored our family tree before. I knew my dad's family was from Florida and that there were some relatives scattered throughout the Deep South. There were rumors about our origins, but that was about it.

So we did one of those DNA tests—full disclosure, this is basically a plug for Ancestry.com—and started building out our family tree. What I discovered was extraordinary.

Tracing the Moye name back from my father, James Morris Moye III, I found a direct line of military service spanning the entire history of American warfare. My dad had served twenty-five years in the Army, including tours in Vietnam during some of the most intense fighting of that conflict. His great-great-grandfather, John E. Moye, had served four years in the Confederate Army during the Civil War, fighting in major battles including Vicksburg. And John's ancestors—George Moye II and his son George Moye III—had both served in the Revolutionary War, fighting in the North Carolina militia from 1775 to 1781.

When I posted about this discovery on LinkedIn—four generations of military service spanning four major American conflicts—it became the most popular post I'd ever written. It resulted in more engagement than anything I'd shared about AI, healthcare, or technology. People were fascinated by this continuous thread of service, this family line that had been present at the founding of our nation and had continued serving through every defining moment of American history.

But what struck me most wasn't just the historical coincidence. It was also the realization that all of these men had faced the same fundamental leadership challenges I'd faced that night over Baghdad, just in different eras and different circumstances.

Captain George Moye II had to convince his neighbors in Pitt County, North Carolina, to risk everything—their lives, their fortunes, and their families—to fight against the most powerful empire in the world. What kind of leadership did it take to sign the Pitt Association in 1775, knowing that if the rebellion failed, signing one's name to that document meant death?

John E. Moye had to maintain unit cohesion and morale through four years of grinding warfare, including the devastating siege of Vicksburg where his regiment suffered 44 percent casualties in a single day at Champion's Hill. How could he keep men fighting when the odds were overwhelming and defeat seemed inevitable?

My father had to lead soldiers through the chaos of the Tet Offensive, coordinate with allied Thai forces, and somehow find time to help rebuild a Buddhist orphanage in the middle of a war zone—because, as he told *Stars and Stripes*, "These are still human beings we're helping."

Different wars. Different weapons. Different enemies. But the same fundamental challenge: How do you lead people through uncertainty, danger, and seemingly impossible odds?

The more I researched these men and their times, the more I realized that the three leadership principles I'd learned that night over Baghdad weren't unique to modern warfare. They were timeless principles that had been tested and proven across generations of American military service.

**Purpose.** When George Moye II gathered his militia company beneath a sycamore tree after drill, he didn't talk about muskets or tactics. He talked about "the right of Pitt to govern Pitt"—connecting his men to something larger than their individual fears.

**Adaptive intelligence.** When John E. Moye's regiment was captured at Vicksburg and later exchanged, they didn't just return to the same old tactics. They adapted: serving coastal defense, guarding prisoners, and eventually joining Sherman's opposition during the Atlanta Campaign.

**Empowered teams.** When my father worked with the Thai Special Liaison Section, he wasn't just following orders from above. He was building relationships, solving problems on the ground, and taking initiative to support an orphanage because he understood the broader mission.

This book is about those timeless leadership lessons and how they apply not just to military service, but also to leadership in any context. Whether you're leading a startup through a critical pivot, managing a team through organizational change, or simply trying to motivate people when morale is low, the principles that guided the Moye family through four generations of American warfare can guide you as well.

This isn't a military history book, though we'll explore enough of each era to understand the context and challenges these men faced. This isn't a genealogy project, though the family connection provides the thread that ties these stories together. This is a leadership book—one that draws on nearly 250 years of tested principles to help you lead more effectively in whatever arena you find yourself.

Because ultimately, whether you're flying over Baghdad in 2003, defending Vicksburg in 1863, fighting at Guilford Courthouse in 1781, or leading a team meeting in 2024, the fundamental challenges of leadership remain the same: connecting people to purpose, adapting to changing circumstances, and empowering others to achieve something greater than themselves.

The stakes may be different, but the principles endure. And sometimes, in our interconnected but often purposeless modern world, we need to look back to move forward—to find in the courage and wisdom of previous generations the tools we need to lead effectively today.

This is their story. And it's a story about leadership that transcends time, place, and circumstance—because the best leadership principles always do.

# *Section I: Purpose*

# CHAPTER 1

# *Purpose over Vision*

## REVOLUTIONARY WAR: PITT COUNTY, NORTH CAROLINA, 1775–1781

IT WAS JULY 1775, just weeks after eighty-eight citizens of Pitt County, North Carolina, had signed what would become known as the Pitt Association—a document pledging their "life and fortune" to whatever course the Continental Congress chose. This amounted to a collective death warrant if the rebellion failed.

George Moye II, a tobacco farmer in his fifties, had signed first. His twenty-three-year-old son, George III, had signed directly beneath his father's name, the ink linking generations like a handoff of command.

Now, as Captain Moye looked out at the faces of Company One—farmers and planters who had traded their hoes for muskets—he faced the same leadership challenge I would encounter 228 years later over the skies of Baghdad: **How do you motivate people to risk everything when the odds seem impossible and the outcome is uncertain?**

The British Empire was the most powerful military force on earth. The Royal Navy ruled the seas. Professional British soldiers had defeated every European power that dared challenge them. And here in eastern North Carolina, along the slow-moving Tar River, a militia captain was asking his neighbors to bet their lives, their families, and their futures on the idea that thirteen loosely connected colonies could defeat the mightiest empire in history.

It was audacious. It was also exactly the kind of moment that reveals what real leadership looks like—and the answer wasn't what most modern leaders might expect.

## THE FRONTIER CONTEXT

To understand Captain Moye's leadership challenge, you need to understand Pitt County in 1775. This wasn't Boston or Philadelphia with their established merchant classes and political networks. Pitt County had been carved out of Beaufort County only fifteen years earlier, in 1760. This was still frontier country—cleared tobacco fields stopping abruptly at walls of swamp cane, plank roads that vanished in spring floods, and fever arriving each May with the mosquitoes.

George Moye II and his wife Elizabeth Gardner had married in neighboring Tyrrell County in 1743—he was barely twenty-one, she was eighteen. They'd moved to claim fifty acres of black loam near Grindle Creek, a tributary of the Tar River. By 1774, they had reared nine children and buried three. Their eldest surviving son, George III, was tall and spare, with calloused palms that testified he knew a hoe as intimately as a musket.

Every dawn before the river fog lifted, father and son walked the tobacco rows. Leaves were wormed by lantern light and suckers pinched before sap could drain the stalk. Payment came months later, often in kind—rum for pork, iron tools for deer hides—because coined money was scarce. Elizabeth kept accounts on pine boards charred with charcoal: "Two barrels meal, four pecks salt, due Captain Moye."

I need to address something here that we cannot ignore: Like most planters of means in colonial North Carolina, the Moyes held enslaved people. This institution—an abomination and permanent stain on our nation—was woven into their world. It doesn't excuse it to say they were creatures of their time, but neither does it erase the leadership lessons their choices under pressure can teach us. What matters for our purposes is what they did when everything was at stake, and how their understanding of purpose evolved through crisis.

This was the practical, grounded world George Moye II knew—where a man's reputation depended on his ability to solve urgent problems with his own hands.

## WHY THEY CHOSE HIM

When county freeholders chose Captain Moye to command Company One of the Pitt militia, they weren't choosing him for pedigree or political connections. They chose him because he embodied what rural Carolina valued: **competence, empathy,** and what we now call **visionary purpose.**

He could survey land, repair wagons, read statute law, and arbitrate quarrels without raising his voice. When drought cracked the furrows, he distributed water from his own cypress cistern. And in local conversations—at church, at the store, at courthouse sessions—he insisted that Pitt's destiny lay not in royal decree but in the honest labor of its citizens.

The county had local bodies called Committees of Safety—essentially judge, jury, and town council rolled into one. They resolved disputes, dealt with taxes, and enforced laws. These committees also maintained small militias, originally formed just to deal with wild animals and the occasional Native American threat.

Captain Moye had earned his rank long before the Revolution began. He drilled his men once each quarter on the courthouse green. But powder was precious, so he ordered blank exercises. His instruction became local legend: "**Sight no higher than a squirrel's head and pull smooth. We will not waste the King's powder on the King's birds.**"

Think about what he was doing here in modern terms: He was building **organizational capability** with minimal resources. He was demonstrating **fiscal discipline** while maintaining readiness. He was creating **muscle memory** that would matter when real bullets started flying.

But more importantly, he was building something that matters more than skills: **collective efficacy**—the shared belief that this group could accomplish difficult things together.

## THE SUMMER OF DECISION

By the summer of 1774, news from the North was filtering down through boatmen and travelers: the Tea Act, the closure of Boston Harbor, and Governor Martin's flight from New Bern. Pitt's Committee of Safety convened repeatedly, counting gun-flints and debating whether discontent warranted open defiance.

This was leadership in an environment of radical uncertainty. No one knew if other colonies would actually fight. No one knew if the Continental Congress would declare independence or seek reconciliation. No one knew if Britain would send overwhelming force or negotiate.

Sound familiar? Whether you're a CEO navigating market disruption, a manager dealing with organizational upheaval, or a team leader facing budget cuts, the fundamental challenge is the same: How do you make decisions and inspire confidence when you can't predict what's coming next?

Captain Moye's answer was to focus on what he could control while preparing for what he couldn't. He increased drilling frequency. He built relationships. He had conversations about what mattered most to his community. And when the moment came to decide, he didn't wait for perfect information.

## JULY 1, 1775: THE MOMENT OF COMMITMENT

Everything changed on July 1, 1775, when eighty-eight citizens gathered in Martinborough (today's Greenville) under a blazing sun to sign the Pitt Association. The document pledged "life and fortune" to whatever course the Continental Congress chose.

But here's what made it brilliant politics and uncertain commitment all at once: The document also professed loyalty to King George III. It was masterful hedging, saying essentially, "We're with Congress, but we're still loyal to the Crown, so please don't burn our barns just yet."

Signing still carried real risk. Britain controlled tobacco exports—these farmers' entire livelihood. Loyalists roamed the backcountry. Barn burners punished dissent with torches. This wasn't an abstract political statement. These men were gambling their families' economic futures.

Captain Moye understood this was fundamentally about economics and autonomy—taxation without representation, the right of people who worked this land to determine their own future. His family's ancestors had come from England just a few generations back. He was asking his neighbors to potentially fight against their own motherland.

He signed first.

Then he handed the pen to his twenty-three-year-old son.

George III could have been excused. His father had influence. He could have found ways to keep his eldest son safe, to protect the family line, to hedge their bets.

Instead, he let his son make his own choice—and by doing so, he demonstrated something crucial about leadership: You can't ask others to take risks that you're not willing to share with those closest to you.

## FROM WORDS TO ACTION

After signing the Pitt Association, Captain Moye didn't just go back to quarterly drills. He doubled the schedule, meeting every other month. By October 12, 1775, his muster showed "58 effectives." George III, who had enlisted as a private, was now a sergeant—learning to lead from his father while carrying his own weight.

After each drill, Captain Moye would gather his men and speak not about muskets or tactics, but about **purpose**. According to records preserved by the Pitt County Historical Society, he would tell them they weren't defending just crops or villages—they were defending "the right of Pitt to govern Pitt."

Let's pause on that phrase: **"The right of Pitt to govern Pitt."**

In eight words, he captured everything. Not abstract concepts such as liberty or independence, but something concrete and personal: the idea that people who worked this land, who knew its rhythms, who had buried their children in its soil, should determine their own future.

This is the difference between **vision and purpose**:

- Vision describes where you're going
- Purpose explains why it matters enough to risk everything

Most leaders articulate vision. They paint pictures of future states, market positions, and organizational transformations. "We're going to be the leading provider of..." "We're going to revolutionize the industry by..." "In five years, we'll have..."

But vision alone doesn't sustain people through crisis. **Purpose does.**

Purpose connects people's daily sacrifices to something larger than themselves but specific enough to feel real. Captain Moye didn't talk about creating a new nation or birthing democracy.

He talked about their community governing itself. He connected immediate action—drilling with unloaded muskets, giving up precious resources, preparing for conflict—to an outcome his neighbors could visualize: their children making their own decisions about their own land.

This is exactly what I had to do over Baghdad when my crew wanted to refuse the search-and-rescue mission. I couldn't motivate them with grand strategy. I had to connect them to something immediate and human: Fellow servicemembers were in danger, and we were the only ones who could help.

The parallel isn't coincidental. Great leaders across centuries understand that **purpose must be both larger than the individual and specific enough to act on.**

## MOORE'S CREEK: FIRST TEST

The first real test came in February 1776. Orders arrived at sunset on February 26: March for Wilmington. Highland Scot Loyalists armed by Governor Martin were advancing to join a British expedition at the coast.

Elizabeth Moye packed jerked venison, cornbread, and two changes of linen for her husband and son. An enslaved teamster named Jacob hitched draft horses to a wagon with spare flints and parched corn. Before dawn, Captain Moye formed Company One.

His words were recorded: "**A cause without sacrifice cannot be claimed.**"

The clash at Moore's Creek Bridge on February 27 lasted only minutes. Patriot artillery tore the Highland charge apart. Loyalist bagpipes fell silent. The victory's effect on morale was seismic—proof that an ordinary militia, properly led, could defeat professional soldiers.

George III ferried prisoners across the creek. His father inventoried captured muskets and powder. When they returned home, Elizabeth wrote letters to neighbors describing the victory, framing it as validation that their choice—risking everything—had been right.

In business terms, this was their **proof of concept**. The team had executed under pressure. The strategy worked. Early wins matter enormously for organizational confidence, especially when you're asking people to commit to something uncertain.

But Captain Moye did something subtle and important in his messaging afterward. He didn't celebrate victory as vindication of his leadership. Instead, he described Moore's Creek as "the gate to our fields"—positioning the win as enabling rather than ending. He told young recruits, "The gate stands open by your hand."

**He gave them ownership of the outcome.**

## THE LONG MIDDLE: 1777–1779

The next three years tested a different kind of leadership. Between 1777 and 1779, there were no major battles in their region, but the war didn't pause—it just changed form.

Continental quartermasters requisitioned wagons and supplies. Loyalist raiders tested ferries and supply lines. Malaria and smallpox decimated ranks as surely as musket fire. Men still had to tend their tobacco crops, maintain their farms, and support their families.

This is the phase most leadership books ignore: **the long middle when nothing dramatic happens, but everything still depends on sustained commitment.**

Think about your own experience. The initial excitement of a new initiative wears off. The daily grind sets in. Resources get stretched. People get tired. Competitors don't cooperate by folding quickly. Market conditions don't improve on your timeline.

How do you maintain commitment during the slog?

Captain Moye's answer: **Never stop demonstrating competence, empathy, and purpose simultaneously.**

He drilled twice monthly, even when men were exhausted. He balanced county ledgers. He presided over the vestry. He showed up consistently, embodying the values he'd asked others to embrace.

Meanwhile, George III was developing his own leadership approach:

- He taught an enslaved blacksmith named Luke to repair gunlocks, boosting local armament by a dozen firelocks monthly—building capability across the organization
- He organized crop-rotation experiments, planting maize after tobacco to restore nitrogen—showing strategic thinking applied beyond the immediate mission
- He courted Sarah Griffin Bradbury, a neighbor's daughter—building alliances and thinking about the future beyond the crisis

In June 1779, when word arrived that British troops had cut inland from Savannah and Pitt detachments were called to Stono Ferry and Brier Creek, casualties were light but morale suffered. The Moyes repeated their creed to anyone who would listen: "**Hold the river. Hold the purpose.**"

## 1780: THE CRUCIBLE

Few Carolina families escaped 1780 unscathed. It was the crucible year—the time when commitment faced its hardest test.

Captain Moye's wagon teams were conscripted by Continental General Benjamin Lincoln, hauling powder kegs through flooded rice fields as British mortars arced over Charleston's bastions. When Lincoln surrendered on May 12, seventeen Pitt militiamen went to prison hulks in the harbor. Jacob, their teamster, disappeared with them. His fate was never recorded.

In August, they followed Horatio Gates to Camden. Hot, malarial, and outnumbered, the militia broke within fifteen minutes. It was a catastrophic defeat.

By chance—some said providence—Captain Moye's company had been detached to guard the wagons in the rear, sparing them heavy casualties. But the commander's conscience took a beating anyway. That night, he wrote in a pocket ledger: **"We must not outrun our purpose; fear fills the void."**

Leadership scholars would later cite this as an early statement of what we now call mission command, but the insight is simpler and more profound: **When you lose sight of purpose, fear rushes in to fill the space.**

This is as true in boardrooms as battlefields. When organizations face setbacks—losing a major account, watching a product fail, or enduring layoffs—the teams that survive are those whose leaders can reconnect them to a purpose even when the immediate situation looks bleak.

Captain Moye couldn't control British troop movements or change the outcome at Camden. But he could control whether his men remembered why they started this fight in the first place. He could anchor them to something that survived the defeat.

## GUILFORD COURTHOUSE: THE THIRD VOLLEY

By winter 1781, General Nathanael Greene had drawn Cornwallis inland, away from coastal supply lines. The stage was set for a confrontation that would define the war in the South.

Captain Moye was in his late fifties now with gout chewing his ankles. He could have stayed home. He'd already given more than most. His son could represent the family.

Instead, he resolved to march.

On March 15, 1781, at Guilford Courthouse, he walked the line of Company One, tapping shoulders with a maple cane. His words were simple: **"Two volleys measured, a third to break them—remember Charleston."**

George III, now a lieutenant, echoed the commands. When British bayonets glittered at fifty paces, Pitt men held fire through two pounding exchanges of musket volleys. Then they loosed a third volley so disciplined that Cornwallis's grenadiers staggered.

A ball grazed George III's left arm, but he caught a wavering private by the collar and barked, "Purpose, lad—purpose!" They fell back in order.

Cornwallis claimed the field, but he'd lost roughly 25 percent of his force. Greene preserved his army intact. Military historians count Guilford among battles "won by losing," but leadership scholars focus on something else: **how losses are managed by anchoring retreat to enduring meaning.**

When George III planted the company banner at the final rally point and shouted, "Pitt still stands!" his men—tattered, bloodied, and defeated in the field—believed him.

Because he'd given them something that survived the immediate loss: proof that their purpose outlasted the day's outcome.

## AFTERMATH: LEADERSHIP THAT HEALS

News of Yorktown trickled south in December 1781. The war was effectively over.

In the first peace months, North Carolina's legislature compensated veterans with land grants. Captain Moye received 300 acres surveyed by his cousin Jesse Moye on the south bank of Grindle Creek.

What he did next reveals the final dimension of leadership purpose.

He divided half the land among his children—ensuring the next generation had a foundation to build on. But he reserved the wettest portion for free Black tenant farmers whose labor he had once owned.

It was rare enough to draw comments in county minutes. And Captain Moye's words were recorded: Leadership "must heal what it once asked men to wound."

It was an imperfect act—nothing could undo the injustice of slavery. But it represented a choice to extend the promise they'd fought for beyond comfortable boundaries. The man who had rallied his neighbors to fight for "the right of Pitt to govern Pitt" was recognizing, however incompletely, that this right had to mean something for all of Pitt's people.

George III took a different path toward reconciliation. He exchanged his half-pay notes for British copper cookware and a battered snare drum captured at Guilford. When he married Sarah Griffin Bradbury in 1796, that drum provided the wedding music—a reminder that survival and purpose could coexist and that the past could inform the future without imprisoning it.

Later, George III served as sheriff of Pitt County, spending considerable energy reconciling with former Loyalists, bringing the community back together. The man who had signed the Pitt Association at age twenty-three spent his middle years healing the divisions that rebellion had created.

## THE LEADERSHIP FRAMEWORK: THREE PILLARS

Looking back across Captain Moye's leadership during the Revolutionary War, three principles emerge that transcend era and circumstance:

### *1. Competence Wins Trust*

Before anyone would follow Captain Moye into rebellion, they had to trust that he could keep them alive. He demonstrated competence in dozens of small ways—drilling without wasting powder, reading statute law, repairing equipment, and managing scarce resources.

**Modern application:** Your team won't commit to your vision if they don't trust your competence. Demonstrate capability in the fundamentals before asking for a commitment on big risks. Show you can manage the details that matter to their daily lives.

### *2. Empathy Wins Hearts*

When drought hit, Captain Moye shared water from his own cistern. When wagons were conscripted, his went too. When it was time to march to Guilford Courthouse despite gout-riddled ankles, he marched.

He didn't ask anyone to take risks he wouldn't share.

**Modern application:** Shared sacrifice builds loyalty that no amount of charisma or compensation can match. When asking teams to work weekends on a critical deadline, leaders should be there too. When announcing budget cuts, leadership should feel the impact first and most.

### *3. Visionary Purpose Binds Both*

Vision describes direction. Purpose explains why the journey matters enough to endure suffering.

Captain Moye could have tried rallying his neighbors with talk of democracy or natural rights or the bright future of a new nation. Instead, he talked about "the right of Pitt to govern Pitt"—connecting their immediate sacrifice to something concrete, personal, and worth protecting: their community, their autonomy, and their children's future.

**Modern application:** When articulating organizational purpose, specificity beats grandeur. "We're going to revolutionize healthcare" is vision. "We're going to help families like yours afford medication that keeps your children alive" is purpose. One inspires speeches. The other inspires sacrifice.

## APPLYING REVOLUTIONARY LEADERSHIP TODAY

The circumstances facing Captain George Moye in 1775 might seem distant from the challenges facing today's business leaders, but the core leadership dynamics are identical:

**You're asking people to commit to uncertain outcomes.** Whether it's pivoting a business model, entering a new market, or restructuring an organization, you're asking people to risk present comfort for future promise. Like the Pitt County farmers, your team is being asked to bet their livelihoods on your judgment.

**You're operating with incomplete information.** Captain Moye didn't know if other colonies would fight, if France would help, or if Britain would send overwhelming force. You don't know if your market assumptions are correct, if your competitor will respond, or if technology will work as planned. Leadership in uncertainty requires demonstrating competence while maintaining flexibility.

**You need sustained commitment through setbacks.** The war didn't end at Moore's Creek. It dragged on for six more years, including catastrophic such defeats as Charleston and Camden. Your initiative won't succeed immediately either. There will be quarters when numbers won't hit, launches will disappoint, and key people will leave. Purpose is what sustains commitment when immediate results disappoint.

**You're building something larger than yourself.** Captain Moye signed the Pitt Association knowing he might not live to see independence. George III took a musket ball at Guilford knowing his father had already given enough for both of them. They fought for something that would outlast them. Great organizational leaders think the same way—building capabilities and cultures that will thrive after they're gone.

## THE THIRD VOLLEY

There's one more lesson from Guilford Courthouse worth carrying forward: the discipline of the third volley.

When British bayonets glittered at fifty paces and every instinct screamed to fire immediately or flee, Captain Moye's training held. His men fired twice, reloaded, and waited for the command to fire the decisive third volley that staggered the British advance.

That third volley—the one that required maximum discipline under maximum pressure—was the one that mattered most.

In business, the "third volley" moments are when you're most tempted to panic, to abandon strategy, to slash costs indiscriminately, and to make reactive decisions. Those are exactly the moments when discipline and purpose matter most.

**The teams that hold their fire, that execute according to plan even when the pressure mounts, and that remember their purpose when fear floods in are the teams that win battles they shouldn't.**

When orders came that February night in 1776, Captain Moye gathered his men and reminded them of what they'd already committed to: "A cause without sacrifice cannot be claimed." Three hours later, they were marching toward their first battle at Moore's Creek Bridge, carrying not just muskets and provisions, but also the unshakeable conviction that what they were fighting for was worth whatever it might cost.

Around 1800, when Captain Moye's heart finally failed, George III carried that captured Guilford drum to the courthouse green, beat three slow rolls, and announced: "My father kept faith—in field and ferry and council hall. Let us keep it, too."

The river still runs black and slow past Grindle Creek, bearing the reflected stars of a nation the Moyes helped imagine, plant, and defend. Their leadership lesson endures: **Give people competence they can trust, empathy they can feel, and purpose they can claim as their own.**

Everything else is just technique.

# CHAPTER 2

# *Purpose Through Catastrophe*

## CIVIL WAR: GEORGIA AND MISSISSIPPI, 1861–1865

On May 16, 1863, Private John E. Moye of the 57th Georgia Infantry watched his regiment get torn apart.

The Battle of Champion's Hill—some called it Baker's Creek—lasted only a few hours, but when Union General Ulysses S. Grant's artillery and infantry finished with them, the 57th Georgia had lost roughly 44 percent of its strength in a single afternoon: forty dead, ninety-six wounded, and forty-eight captured. The survivors fell back in chaos toward Vicksburg, leaving their dead where they'd fallen on a Mississippi hillside thirteen hundred miles from home.

John Moye, somehow, wasn't among the casualties. But as he stumbled back through the dust and smoke, he faced a question that every leader eventually confronts: **When the mission is failing, when the losses are catastrophic, and when continuing forward seems pointless, how do you find the purpose to keep going?**

This wasn't his first setback, and it wouldn't be his last. Over the next six weeks, he would endure the Siege of Vicksburg—trapped in a fortress city while Union artillery pounded them daily and supplies dwindled to nothing. On July 4, 1863, he would surrender along with the remaining 342 men of his regiment and become a prisoner of war.

For most people, the war would have ended there. But John Moye's war didn't. After being paroled and exchanged, he rejoined the fight. He would go on to guard prisoners at Andersonville, resist Sherman's advance through Georgia, march with Hood into Tennessee, and finally surrender—again—in April 1865 as part of the last Confederate army still in the field.

**He fought from the first volunteer call in October 1861 until the final surrender in April 1865. He saw his cause defeated, his regiment destroyed, his homeland burned, and his way of life ended. And yet he kept going.**

The question isn't whether his cause was just—it wasn't. The Confederacy fought to preserve slavery, and that makes it impossible to romanticize anything about their struggle. But leadership lessons aren't about politics or moral righteousness. They're about how humans maintain commitment under extreme duress, how teams survive catastrophic failure, and how individuals find purpose when all the rational arguments say to quit.

Because if you lead long enough, you will face your own Champion's Hill. You will watch something you built get dismantled. You will see a market shift destroy your business model. You will endure a quarter that erases years of progress. And the question will be the same one John Moye faced: **Do you give up, or do you find a way to keep your people moving forward?**

## FROM REVOLUTION TO REBELLION

The path from George Moye III's Revolutionary War service to John E. Moye's Confederate service runs through the red clay of Georgia cotton country.

After the Revolutionary War ended, North Carolina compensated its veterans with land grants—often in newly opened territories including Georgia. In 1802, George Moye III took his family from Pitt County, North Carolina, to Washington County, Georgia, where cotton was replacing tobacco as the cash crop of choice.

He brought his enslaved laborers with him. This brutal institution that his father had begun to question in his final years became the economic foundation of the family's new Georgia life. George III and his wife Sarah had multiple children, including a son named Durin Griffin Moye, born in 1799.

Durin married Mary Penelope Page around 1829. She was the daughter of Solomon Page, another Revolutionary War veteran. Together they built a substantial plantation in Davisboro, Washington County. Their first child, born June 1, 1825, was John E. Moye.

By the time John reached adulthood, the Moye family had transitioned from small-scale North Carolina tobacco farmers to Georgia cotton planters with considerable holdings—and considerable dependence on enslaved labor. When John married and started his own family, he continued the pattern, establishing himself as a planter in his own right.

This matters for understanding his decision to enlist: **John Moye wasn't fighting for abstract principles. He was fighting to preserve an entire economic and social system that his family had built over three generations.**

When South Carolina seceded in December 1860 and Georgia followed in January 1861, John was thirty-five years old—a husband, a father, a man with property and prospects. He could have stayed home. At thirty-five, with farming responsibilities, he wasn't young and foolish. He was making a calculated choice about what mattered most.

On October 12, 1861, he enlisted as a volunteer in what would become the 57th Georgia Infantry.

## THE REGIMENT

The 57th Georgia Infantry was organized in the spring of 1862 from volunteer companies drawn from central Georgia counties—Troup, Peach, Montgomery, and Oconee. It wasn't one of the storied regiments that had fought at First Manassas or gained fame in early battles. It was a workmanlike unit of farmers and tradesmen.

John Moye served in Company G. Interestingly, at least two other soldiers surnamed Moye served in the regiment: cousins Corporal George J. Moye and Private Virgil M. Moye. The Moye family was a tight group; men from the same region enlisting together, looking out for each other.

The 57th spent most of 1862 away from Georgia entirely, deployed to east Tennessee and Kentucky as part of the Confederate Heartland Offensive. Generals Edmund Kirby Smith and Braxton Bragg launched a campaign into Kentucky in late summer 1862, hoping to secure that border state for the Confederacy. The 57th Georgia spent months on garrison duty, marches, and minor skirmishes—learning the tedium of military life but seeing little major combat.

It was, in retrospect, a grace period before the real war found them.

## CHAMPION'S HILL: WHEN STRATEGY FAILS CATASTROPHICALLY

In early 1863, the 57th Georgia was transferred to Mississippi and assigned to Brigadier General Alfred Cumming's Georgia Brigade, part of the force defending Vicksburg.

By spring 1863, Union General Ulysses S. Grant had become obsessed with taking Vicksburg. The city sat on high bluffs overlooking the Mississippi River, controlling north-south river traffic. If Grant could take it, he would split the Confederacy in two, cutting Texas, Louisiana, and Arkansas off from the rest of the South.

Grant tried multiple approaches—through canals, over bayou routes, and by direct assaults—before finally executing a brilliant and risky maneuver: marching his army down the Louisiana side of the river, crossing below Vicksburg, and attacking from the landward side. It was a campaign of speed and aggression, catching Confederate commanders off guard.

On May 16, 1863, Grant's forces caught up with the Confederate army withdrawing toward Vicksburg at Champion's Hill, about twenty miles east of the city.

The 57th Georgia, part of Cumming's brigade, held a position on the Confederate left. When Grant's assault came, it came hard. Union artillery pounded the Georgia regiments. Infantry charged up the hill. Hand-to-hand combat erupted across the position.

Contemporary accounts describe chaos—units breaking, officers trying to rally their men, the wounded crying out, and the dead piling up. The 57th Georgia soldiers stood for as long as they could, but when the Union breakthrough came, they had no choice but to fall back.

**When the butcher's bill came due, the 57th Georgia had lost roughly 44 percent of its strength in one afternoon.**

Think about that for a moment. Nearly half your team—gone. Not transferred; not quit; but dead, wounded, or captured. And you're not going home. You're falling back to continue the fight with whoever's left.

John Moye's name doesn't appear in the casualty lists from Champion's Hill, which means he survived. But survival came with its own burden: **the responsibility to keep going when so many others couldn't.**

## THE SIEGE: ENDURING THE UNENDURABLE

The survivors of Champion's Hill retreated into Vicksburg's defensive perimeter. The city sat on high bluffs above the Mississippi

River, naturally defensible. Confederate engineers had constructed a nine-mile arc of trenches, redoubts, and artillery positions around the landward side.

Grant's army arrived and encircled the city. On May 22, Union forces launched a massive assault—and were bloodily repulsed. Confederate defenders proved that the fortifications were too strong for a direct attack.

So Grant settled in for a siege.

For the next forty-seven days, the defenders of Vicksburg endured systematic destruction. Union artillery fired day and night. Naval gunboats on the river added their shells. Supplies dwindled. Soldiers and civilians alike dug caves into the hillsides to escape the bombardment.

Rations were cut and then cut again. Men went hungry. Disease spread in the crowded, unsanitary conditions. The constant bombardment shredded nerves. There was no relief, no resupply, and no possibility of escape.

**This is the leadership challenge that no one prepares you for: holding your team together when everyone knows the situation is hopeless.**

Inside Vicksburg, everyone understood the math. No Confederate army was coming to break the siege. No supplies were getting through. Eventually, starvation would force surrender. It was only a question of when.

Modern business analogies feel inadequate for this level of suffering, but the psychological dynamics aren't that different. Think about leading a division that's being shut down. Corporate has made the decision. Everyone knows it's coming. But you still have to keep people showing up, doing the work, and maintaining standards for weeks or months until the official end date.

Or consider leading a startup that's running out of runway. The investors have stopped returning calls. The pivot isn't working. Everyone can see the burn rate and can do the math.

But you still need the engineering team to keep building, the sales team to keep calling, and everyone to keep executing until the very last day.

**The question becomes: What do you tell them? How do you maintain any sense of purpose when the outcome is predetermined?**

We don't have John Moye's personal writings from inside Vicksburg. But we know he was there for the full forty-seven days, and we know he was still with his unit when the end came.

On July 4, 1863, Confederate General John C. Pemberton surrendered Vicksburg. The remaining 342 men of the 57th Georgia Infantry—down from over 600 who'd started the campaign—marched out and stacked their arms. They became prisoners of war on Independence Day.

Grant, in a gesture of practical generosity, paroled most of them. They were released on their promise not to fight until officially exchanged—meaning until a Confederate prisoner held by the Union was swapped for them, releasing them from their parole obligation.

**For most men, the war would have ended there. They could go home, tend their farms, be with their families, and claim they'd fulfilled their obligation.**

John Moye made a different choice.

## THE DECISION TO RETURN

Sometime in late summer or fall of 1863, John Moye was officially exchanged. The paperwork processed. He was legally free to rejoin Confederate forces.

And he did.

Think about that decision for a moment. You've been through Champion's Hill. You've endured the Siege of Vicksburg. You've been captured, imprisoned, and paroled. You're in your late thirties. You have a wife and children back in Georgia. You could go home.

**Why go back?**

This is where purpose transcends rational calculation. By late 1863, the Confederacy's strategic situation had deteriorated dramatically. Gettysburg had failed. Vicksburg had fallen. The Union controlled the Mississippi. The blockade was tightening. Any objective observer could see the trajectory.

But John Moye—and thousands like him—went back anyway.

Not because they thought they would win. Not because the odds looked good. But because **purpose isn't always about winning. Sometimes it's about who you are, what you believe you're fighting for, and what you can't walk away from, even when logic says you should.**

I want to be crystal clear: The cause he fought for was morally indefensible. The Confederacy fought to preserve slavery, and no amount of talk about states' rights or heritage or economic systems changes that fundamental evil. But we're not examining whether his purpose was just. We're examining how purpose—even misguided purpose—sustains commitment through catastrophic failure.

Because here's the uncomfortable truth: **Good leaders with bad causes can still teach us about leadership. The Confederacy was wrong, but Confederate soldiers' commitment to their units, their comrades, and their communities offers lessons about human motivation that transcend politics.**

## BACK TO THE FIGHT

After his exchange, John Moye didn't immediately rejoin the main Confederate army. The 57th Georgia, shattered at Vicksburg, was sent back to Georgia for rest and refit. By the end of 1863, they were assigned to coastal defense at Savannah.

In February 1864, they fought a skirmish at Whitemarsh Island near Savannah—nothing major, but just repelling a Union probing attack. Then came an assignment that must have tested every man's spirit: guarding Union prisoners at Andersonville Prison in spring 1864.

Andersonville would become infamous as one of the war's worst prison camps, where thousands of Union soldiers died from disease, exposure, and malnutrition. The Confederate guards themselves were poorly supplied and sick. It was misery guarding misery.

By mid-1864, the 57th Georgia was pulled from prison duty and sent to rejoin the Army of Tennessee, now fighting desperately to stop Sherman's advance through Georgia toward Atlanta.

John Moye participated in the Atlanta Campaign battles: Kennesaw Mountain in June, where Confederate troops held a strong defensive position; Peachtree Creek and the Battle of Atlanta in July, where desperate Confederate counterattacks failed to stop Sherman's grinding advance; and Jonesboro in late August, where the final Confederate effort to save Atlanta collapsed.

Atlanta fell on September 2, 1864.

**The 57th Georgia had now watched their cause lose its most important city, its major supply hub, and any realistic hope of victory.**

And still they kept fighting.

General John Bell Hood took what remained of the Army of Tennessee and marched north into Tennessee in a desperate and ultimately futile attempt to draw Sherman away from his March to the Sea. The 57th Georgia marched with them through late 1864.

When that campaign collapsed in blood and failure, the remnants of thc army retreated east to the Carolinas in early 1865, trying to join forces with other Confederate units to somehow stop Sherman's juggernaut.

## THE FINAL SURRENDER

By April 1865, the 57th Georgia Infantry had been so reduced that it was consolidated into the 1st Georgia Consolidated Infantry—combining the remnants of multiple shattered regiments into a single functional unit.

On April 26, 1865, that consolidated unit surrendered at Greensboro, North Carolina, as part of General Joseph E. Johnston's army—one of the last Confederate armies still in the field.

John E. Moye had served from October 12, 1861, until April 26, 1865—three and a half years, from hopeful volunteer to bitter survivor; from Champion's Hill to Vicksburg to Andersonville to Atlanta to Tennessee to the Carolinas.

**He fought from the beginning to the end, through every defeat, through capture and release, and through the disintegration of everything he was fighting for.**

When he finally went home to Georgia—to Washington County if his property still stood, or to Laurens County if he had to relocate—he found a world transformed. Slavery was abolished. The plantation economy was destroyed. The social order had been upended. Everything he'd fought to preserve was gone.

He lived another forty-one years, dying in 1906 at age eighty-one. His son Robert George Washington Moye—my great-great-grandfather—would carry the family line forward, adapting to the new South and eventually moving to Florida where my father's branch of the family would settle.

But John E. Moye's war never really ended. How could it? He'd given four years, lost comrades, endured capture, watched his regiment decimated multiple times, and seen his cause utterly defeated.

**The question isn't whether he should have quit earlier. The question is: What can we learn from someone who maintained**

**commitment through that level of sustained catastrophic failure?**

## THE DARK SIDE OF PURPOSE

Before we extract leadership lessons from John Moye's endurance, we need to confront an uncomfortable reality: **Purpose can sustain people through suffering for causes that are morally wrong.**

The same psychological and organizational dynamics that help good leaders keep good teams committed to worthy goals also helped Confederate leaders keep Confederate soldiers fighting for slavery. Purpose is morally neutral. It's a tool, and like any tool, it can be used for good or ill.

This is why modern leaders must constantly examine not just whether their teams are committed, but whether they're committed to something worth the commitment. John Moye's regiment kept fighting long after rational analysis said the cause was lost—but the cause wasn't worth fighting for in the first place.

**Endurance without moral purpose is just stubbornness in service of the wrong thing.**

That said, if we look at John Moye's individual motivations—setting aside the Confederacy's cause—we can identify dynamics that apply to any leader trying to sustain commitment through adversity:

### *Lesson 1: Purpose Rooted in Community Outlasts Strategic Setbacks*

John Moye didn't enlist to defend Jefferson Davis's government or constitutional theories about states' rights. He enlisted because his community, his family's livelihood, and his way of life were under threat.

When Champion's Hill failed, when Vicksburg fell, and when Atlanta burned, those strategic defeats didn't erase his community bonds. The men of Company G were still his neighbors. The fields of Washington County were still his home. His family's future was still at stake.

**Modern application:** When your company loses a major account, when your product launch fails, or when the quarterly numbers disappoint, your team members can lose faith in the strategy. But if their commitment is rooted in bonds to each other and belief in the team's capabilities—not just the current plan—they'll keep fighting while you develop the next strategy.

This is why companies that build strong cultures survive strategic pivots. The community bonds outlast the strategic setbacks.

### *Lesson 2: Shared Suffering Creates Powerful Bonds*

The 57th Georgia Infantry lost 44 percent of its strength at Champion's Hill and then endured forty-seven days of siege. Those shared experiences created bonds that no amount of defeat could break.

Research on military units consistently shows that soldiers don't fight primarily for ideology or grand strategy. They fight for the men next to them. They endure because their brothers-in-arms are enduring. Shared suffering creates powerful commitments.

**Modern application:** Teams that survive crises together develop loyalty that can't be bought or manufactured. The startup that barely survived its first year develops cohesion that established companies envy. The division that weathered layoffs together has bonds that transcend individual ambition.

Great leaders understand this, and when possible, they keep teams together through adversity rather than constantly reshuffling people to avoid association with failure.

### *Lesson 3: Small Wins Matter Even When the Big Picture Is Failing*

The 57th Georgia didn't win major strategic victories after Vicksburg, but they won small ones: holding their position at Kennesaw Mountain for another day, repelling an attack at Whitemarsh Island, and maintaining unit cohesion during retreats.

**When you can't win the war, you focus on winning the battles. When you can't win the battles, you focus on winning the day. When you can't win the day, you focus on maintaining discipline and unit cohesion.**

This cascading scale of purpose—from strategic to operational to tactical to personal—is how effective leaders sustain commitment through losing campaigns.

**Modern application:** When your company is being acquired and shut down, you focus on getting severance packages for your team. When the product is being discontinued, you focus on transitioning customers smoothly. When the project is canceled, you focus on capturing lessons learned and maintaining your team's reputation.

You find the smallest level at which you can still "win" something, and you give your people purpose at that level.

### *Lesson 4: The Option to Quit Makes Staying Meaningful*

After Vicksburg, John Moye could have gone home. He was paroled. He'd fulfilled his obligation. No one would have called him a coward.

**The fact that he chose to return gave his continued service meaning.**

This is paradoxically how voluntary commitment works: **The option to quit makes staying powerful.** If people have no choice, endurance is just submission. If they have a choice and choose to stay, endurance becomes a statement of purpose.

**Modern application:** The most committed teams are those whose members could work anywhere but choose to stay. The most loyal employees are those who could get recruited away but don't. The most powerful organizational cultures are those where people stay by choice, not necessity.

This is why retention of top talent matters so much—it's not just about keeping their skills; it's also about the message their choice to stay sends to everyone else.

### *Lesson 5: Purpose Requires Honest Reckoning Eventually*

John Moye survived the war and lived another forty-one years in the reconstructed South. We don't have his personal writings, but we know he had to reconcile what he'd fought for with what actually happened.

The South lost. Slavery ended. The plantation economy he'd fought to preserve was destroyed. He had to adapt to a world where his original purpose—preserving his family's economic and social position—no longer mapped to reality.

**This is the leadership challenge that John Moye faced after April 1865, and it's one that modern leaders face too: What do you do when your purpose becomes obsolete?**

**Modern application:** When your company's core product becomes technologically obsolete, when your industry gets disrupted, or when the market moves away from what you do best—clinging to the old purpose becomes destructive. Great leaders help their teams find new purpose while honoring what they endured for the old one.

This is the difference between organizations that die clinging to the past and those that successfully reinvent themselves. The latter acknowledge the commitment people made to the old purpose while articulating a new one worth fighting for.

## THE UNCOMFORTABLE TRUTH

John E. Moye's story makes us uncomfortable because it demonstrates that commitment, endurance, and purpose—qualities we admire—can be directed toward causes that are fundamentally wrong.

**This is why leadership divorced from moral purpose is dangerous.**

The same organizational dynamics that made the 57th Georgia Infantry continue fighting through defeat after defeat could make a corporate team continue pursuing a harmful product, a fraudulent business model, or an unethical strategy.

Consider Elizabeth Holmes at Theranos. She articulated a compelling purpose—revolutionizing healthcare through a single drop of blood, saving lives through breakthrough technology. Her team worked grueling hours, maintained secrecy, and pushed through mounting doubts because they believed in that purpose. Yet without moral grounding, that purpose became a shield for fraud. Patients received faulty test results. Investors lost billions. The compelling vision obscured the ethical rot beneath.

Or Sam Bankman-Fried at FTX, who spoke eloquently about "liberating" people through cryptocurrency, creating a new kind of economy free from traditional financial systems' constraints. His employees worked around the clock, convinced they were building something revolutionary. That sense of purpose kept them loyal even as customer funds were misappropriated and basic financial controls ignored. Purpose without integrity became purpose weaponized.

True leadership requires more than a compelling purpose. It requires that the purpose be tethered to enduring moral principles—honesty, human dignity, and justice. Without that anchor, the very qualities that make organizations resilient can make them destructive. Purpose sustains commitment. But **leaders**

**bear the responsibility for ensuring that the purpose is worth sustaining.**

John E. Moye fought for a "way of life" that included the enslavement of human beings. His purpose—protecting his family, his land, and his community—was real. His courage was real. His loyalty to his comrades was real. But courage, loyalty, and purpose in service of an immoral cause remain immoral, no matter how sincerely held or bravely defended. The Confederacy's purpose—preserving slavery and white supremacy—was morally indefensible. His endurance, his commitment, and his willingness to return to the fight after Vicksburg demonstrate powerful psychological dynamics, but they don't validate his cause.

**The leadership lesson isn't "Be like John Moye." It's "Understand the dynamics that kept John Moye fighting, and direct those dynamics toward purposes that make the world better, not worse."**

## FINDING PURPOSE AFTER CATASTROPHE

One more aspect of John Moye's story is worth examining: **How do you rebuild after total defeat?**

When John Moye returned to Georgia in 1865, he faced a world turned upside down. The Confederate currency was worthless. Much of Georgia's infrastructure had been destroyed by Sherman's army. The labor system his family's plantation had depended on was abolished. The social order was in chaos.

He could have become bitter, clinging to the lost cause, spending the rest of his life refighting the war in his head. Many Confederate veterans did exactly that, joining organizations that mythologized the Old South and opposed Reconstruction.

We don't know which path John Moye took. But we know he survived, rebuilt some kind of life, and passed the family line forward to the next generation. His son Robert George Washington Moye, born in 1852, would live until 1930—spanning the entire period from antebellum South through Reconstruction, the Gilded Age, and into the modern era.

**Modern application:** Sometimes the purpose you've committed to fails completely. The startup shuts down. The division gets sold. The company goes bankrupt. The project is canceled and will never see the light of day.

In those moments, great leaders help their teams find meaning in what they accomplished, grieve what was lost, and then articulate a new purpose worth pursuing. They don't pretend the failure didn't happen or that the suffering was wasted. They acknowledge both and then help people move forward.

## FROM CHAMPION'S HILL TO YOUR CRISIS

John E. Moye stood on Champion's Hill on May 16, 1863, and watched his regiment get destroyed. Then he endured forty-seven days of siege, surrendered, was paroled, chose to return, and fought for nearly two more years until the final surrender.

**That's not rational. That's purpose.**

The next time your strategy fails catastrophically—the market turns, the product flops, the acquisition falls through, or the funding dries up—you'll face a version of the same question John Moye faced: What do you tell your team? How do you find purpose to keep going when all the rational arguments say quit?

The answer isn't to blindly endure. John Moye's example also shows the danger of commitment to a cause that's already lost. Sometimes the right answer is to surrender, to pivot, to admit defeat, and to move on.

But if you're going to ask your team to stay in the fight through adversity, you need to understand the kind of purpose that sustains people through suffering:

- Purpose rooted in community bonds, not just strategic plans
- Purpose built through shared experience, especially shared hardship
- Purpose scaled appropriately to what you can actually control
- Purpose chosen freely, not imposed or assumed
- Purpose examined constantly to ensure it remains worth the commitment

These dynamics kept the 57th Georgia Infantry together through four years of war. They've kept companies together through near-bankruptcy. They've kept teams together through restructurings, pivots, and crises.

**Purpose is the answer to catastrophe—but only if the purpose is worth having in the first place.**

John E. Moye died in 1906, forty-one years after Appomattox, in a world utterly different from the one he'd fought to preserve. Whatever purpose he found for those four decades after the war, it had to be something new—built on the ruins of the old but facing forward, not back.

That's ultimately the leadership lesson from his Civil War service: **Purpose sustains you through catastrophe, but wisdom requires knowing when to find new purpose from the ashes of the old.**

# CHAPTER 3

# *Mission First, People Always*

## VIETNAM WAR: CAMP BEARCAT, 1967–1968

IN EARLY 1968, NOT long after the Tet Offensive had turned Saigon's streets into battlefields, my father, Captain James "Jim" Morris Moye III, learned about an orphanage.

He was serving as the civil affairs officer for Camp Bearcat, a sprawling U.S. military base in Bien Hoa Province about twenty miles southeast of Saigon. His job involved coordinating with local Vietnamese officials, managing civic action programs, and generally trying to win hearts and minds in a war zone where "winning" increasingly seemed like an abstraction.

Someone mentioned that Buddhist monks were caring for war orphans in an isolated village just three thousand meters from Bearcat's perimeter—close enough to see, and dangerous enough that most Americans avoided the area.

Captain Moye went looking for it. What he found was about seventy children living with monks and nuns in conditions of extreme austerity: huts with leaky roofs, no clean water, no medical supplies, and barely any food. The monks were trying to build a self-sufficient farm and shelter with almost no resources, caring for children whose parents had been killed in a war that showed no signs of ending.

When Captain Moye asked how his unit could help, the head monk, Mr. Nguyen Van Siu, told him something that stuck: **"We don't want your money. Money brings evil."**

But they needed help. Desperately. Over the next three months, Captain Moye arranged for:

- Medical teams (MEDCAPs) to treat the children
- Surplus building materials from base supplies
- Engineers to help with construction
- A 72-foot water well dug by hand
- Privately donated medical supplies funneled through military channels

Within three months, the orphanage had new buildings, fresh water, and a working dispensary. When *Stars and Stripes* interviewed him about the project, Captain Moye said: **"We heard there was an orphanage in the area and we went looking for it. We found a lot of very determined people just 3,000 meters from our perimeter."**

And later: **"These are still people."**

That phrase—"these are still people"—captures everything about leadership in complex, morally ambiguous situations. Captain Moye was fighting a war. His battalion was engaged in combat operations. Tet had just proven that nowhere was safe, and that even Saigon could be overrun. The idea that he should be diverting resources, time, and attention to a Buddhist orphanage outside his perimeter might have seemed to be a distraction from the mission.

But Captain Moye understood something that many leaders miss: **Caring for people isn't a distraction from the mission. It IS the mission.**

This is the leadership lesson from his Vietnam service: **Mission first, people always.** The two aren't in conflict. The best leaders accomplish both simultaneously.

## THE PATH TO VIETNAM

James Morris Moye III was born April 19, 1940, in Sanford, Florida—a backwater town in what is now in the Orlando area but was then just swamp, citrus groves, and hardscrabble living. His father, James Morris Moye Jr. (they called him "Mutt"), worked as a railroad engineer. His mother, Helen Wilson, managed a household with four children: two older daughters, James, and a younger son.

It wasn't an easy childhood. Mutt was an alcoholic and hard on the family. Money was tight. Opportunities were limited. When James graduated from high school in 1958, he saw the military as a way out—a path to something better than what Sanford, Florida, offered.

(His father, incidentally, had missed both World Wars—he was ten years old when World War I started, thirty-five when World War II began, although he did put his name in the draft. Timing matters in military service: James Morris Moye Jr. sat out both conflicts not by choice but by chronology.)

Young James enlisted in the Army for a three- or four-year term in the early 1960s. It was peacetime service, nothing dramatic. But while stationed at Fort Bliss in El Paso, Texas, he met a college student named Kathy Oth. When he asked her to marry him, her father had one requirement: **James needed to demonstrate he could support a family.**

So in 1964, James re-enlisted. They married in August 1964.

Their first son, James Morris Moye IV, was born June 9, 1966, in Germany, where James was stationed. In 1967, the family moved back to El Paso. And in late 1967, with Kathy pregnant with their second son, James received orders for Vietnam.

His second son, Mike, was born in 1968 while Jim was deployed. The Tet Offensive erupted while his wife was giving birth back in Texas. Jim Moye wouldn't meet his second son for months.

This was the reality of Vietnam service: **You deployed, left your family behind, and hoped you'd come back to meet your children.**

## THE 9TH INFANTRY DIVISION AND CAMP BEARCAT

Captain Moye was assigned to Headquarters and Headquarters Company (HHC), 2nd Battalion, 47th Infantry Regiment, 9th Infantry Division. The 9th Infantry Division—nicknamed the "Old Reliables"—had been reactivated specifically for Vietnam service, reorganized for counterinsurgency operations in the Mekong Delta and III Corps regions around Saigon.

The division began deploying in late 1966. By January 1967, they'd established their main base at Camp Bearcat (also known as Camp Martin Cox) in Long Thanh, Bien Hoa Province. Bearcat was part of the "ring" of U.S. divisions protecting Saigon from communist forces operating in the surrounding provinces.

The 2/47th Infantry was a mechanized battalion—meaning they operated M113 armored personnel carriers. These tracked vehicles provided mobility and firepower in the Delta's difficult terrain and could serve as mobile platforms for infantry operations.

Throughout 1967, Captain Moye's battalion participated in major operations:

**Operation Junction City (February–March 1967)** was a massive multi-division search-and-destroy operation in War Zone C, Tay Ninh Province, aimed at finding and eliminating COSVN—the Central Office for South Vietnam, the communist headquarters for operations in the South. The 9th Division saw its first major pitched battles during Junction City. On March 20, 1967, a mechanized cavalry troop attached to the 9th Division was ambushed near Bau Bang, resulting in 230 enemy killed for 4 U.S. casualties—the kind of lopsided engagement that looked good in reports but didn't change the strategic situation.

**Operations Enterprise and Akron** involved the 9th Division conducting sweeps through Long An Province and adjacent areas, trying to interdict Viet Cong supply lines and strongholds. The 2/47th Infantry often served as mobile reserve, using their M113s to respond quickly along Route 15 (nicknamed "Thunder Road").

**Operations Coronado** in mid- to late-1967 saw the 9th Division's 2nd Brigade form the Army component of the Mobile Riverine Force, working with Navy Task Force 117 to conduct operations in the Mekong Delta's waterways. Coronado VIII in late September 1967 took place in the ominously nicknamed "Forest of Assassins" and marked the first time new Thai allied forces joined U.S. operations.

By late 1967, the 9th Division was shifting its center of gravity southward to the Delta while maintaining presence at Bearcat. This required creating Task Force Forsyth, a provisional command built around the 2/47th Infantry to hold Bearcat and the surrounding operational area.

Captain Moye, as part of HHC, would have been involved in the battalion staff work—operations planning, coordination with other units, and increasingly, civil affairs and liaison work with allied forces.

## THE THAI CONNECTION

One of the most interesting aspects of Captain Moye's service was his work with allied Thai forces.

In September 1967, the Royal Thai Army Volunteer Regiment—nicknamed the "Queen's Cobras"—arrived in Vietnam: 2,200 Thai soldiers who based themselves at Camp Bearcat alongside the 9th Division. Thailand was one of the "Free World Military Forces" contributing troops to support South Vietnam, and the Queen's Cobras were their initial deployment.

The Thai forces focused on pacification missions in Bien Hoa Province, conducting sweeps through the Nhon Trach jungle areas southeast of Saigon. They also undertook civic action projects—building roads, establishing a local hospital, and winning hearts and minds through development work alongside combat operations.

Captain Moye, who learned to speak some Vietnamese during his deployment, worked closely with the Thai contingent. In March 1968, II Field Force created a Special Liaison Section (SLS)—about eighteen U.S. personnel tasked with coordinating American support for Thai operations. The SLS handled artillery support, air support, medical evacuation, and logistics for the Thai forces, essentially serving as the communication bridge between Thai commanders and U.S. assets.

While it's not entirely clear whether Captain Moye was formally assigned to the SLS or simply worked closely with it as Bearcat's civil affairs officer, his role definitely involved Thai coordination. When the much larger Thai Black Panther Division arrived in mid-1968 (about 11,000 troops), they took over Camp Bearcat entirely as their headquarters, and the liaison work became even more critical.

**Modern business application:** Captain Moye was managing a complex, multi-national partnership under combat conditions. He was coordinating logistics, managing cultural differences, and aligning objectives between organizations with different command structures and priorities. This is fundamentally the same challenge facing any leader managing strategic partnerships, joint ventures, or cross-functional teams.

The lesson: **Effective partnership requires someone who speaks both languages (literally and figuratively) and can translate between organizational cultures.**

## TET: JANUARY 31, 1968

The night of January 31, 1968, changed everything.

The Tet Offensive—a coordinated communist assault on cities and bases across South Vietnam—erupted during the Lunar New Year celebrations. Seventy thousand Viet Cong and North Vietnamese Army troops attacked more than a hundred cities, towns, and bases simultaneously.

Camp Bearcat itself didn't face a major direct assault, but the 2/47th Infantry was immediately thrown into the fight. They rushed to Long Binh Post, about ten miles away—the massive U.S. logistics base and II Field Force headquarters. Viet Cong sappers had penetrated the perimeter and blown up portions of the ammunition depot. The 2/47th's mechanized infantry helped defend the II Field Force command post and secure the critical supply areas.

Then, almost immediately, they were redirected into Saigon itself.

For four days (February 1–4), Captain Moye's battalion fought in the streets of Cholon, the dense urban district in southern Saigon near the Phu Tho racetrack. The M113 armored personnel carriers—designed for rice paddies and jungle trails—proved surprisingly effective in urban combat. Their heavy machine guns suppressed sniper fire, and their armor protected infantry clearing buildings.

**Think about the leadership challenge here:** Captain Moye and his colleagues had trained for riverine operations and rural counterinsurgency. Now they were fighting house-to-house in a city. They had to adapt tactics, learn urban combat on the fly, and keep their troops focused when the entire strategic picture of the war had just been upended by a nationwide enemy offensive.

By the end of Tet's first week, the 9th Division had killed over 1,600 Viet Cong and NVA troops and helped secure Saigon,

Long Binh, and other critical locations. Tactically, it was a victory. Strategically and psychologically, Tet was a disaster for the U.S.—American public support for the war plummeted when they saw that after years of "light at the end of the tunnel" rhetoric, the enemy could still attack anywhere, anytime.

Then in May 1968 came "Mini-Tet"—a second wave of attacks. On May 7, 1968, the 2/47th rushed from Bearcat to Saigon's Eighth District to reinforce U.S. military police and ARVN forces battling several Viet Cong battalions at Cau Mat hamlet and the Y-Bridge area. Company B led the charge and hit fierce resistance—rockets, automatic weapons, and ambushes in dense urban terrain.

The mechanized troops fought through the night in intense house-to-house combat. The battalion commander had to arrange emergency ammunition resupply from Bearcat because they'd expended so much ordnance. Eight men from the 2/47th were killed in this action. By morning, they'd cleared Cau Mat and secured the Y-Bridge, but it was another costly reminder that this war had no secure rear areas.

**Leadership lesson:** When the situation changes overnight—when your industry gets disrupted, when the market shifts, or when a competitor makes an unexpected move—the teams that survive are those whose leaders can help them adapt quickly while they maintain unit cohesion. Captain Moye couldn't control the strategic situation or predict communist offensives. But he could ensure his people stayed focused on their immediate mission, supported each other, and executed professionally under extreme stress.

## THE ORPHANAGE: PURPOSE BEYOND THE MISSION

It was in this context—after Tet, in the middle of a war that seemed increasingly pointless to many Americans but still demanded everything from the troops on the ground—that Captain Moye learned about the orphanage.

He could have ignored it. He had plenty of justification: combat operations, security concerns, resource constraints, and bureaucratic hurdles. There were a thousand reasons why helping seventy orphans three thousand meters outside the wire wasn't his problem. Instead, he went looking for it.

When he found the orphanage, he discovered Mr. Nguyen Van Siu and the other monks caring for children in conditions of desperate poverty. The monks were suspicious at first—why would American soldiers help? Were they looking for intelligence? Did they want something in return?

Captain Moye, through his rudimentary Vietnamese and interpreters, explained: **"We're here. You need help. We can provide it."**

The monk's response—"We don't want your money; money brings evil"—forced Captain Moye to think creatively. Money wasn't the answer. But his base had surplus building materials. His unit had engineers who could help dig a well. The medical corps could provide checkups for the children. Supply channels could be used to funnel donated medical supplies.

Over three months, the transformation was remarkable. New buildings went up. A 72-foot well provided clean water. Children received medical care. The dispensary got stocked with basic supplies.

**And here's what makes this a leadership lesson rather than just a heartwarming story: Captain Moye didn't do this instead of his job. He did it as part of his job.**

He understood that engaging in civil affairs, winning hearts and minds, and demonstrating that Americans cared about Vietnamese lives weren't distractions from combat operations—they were essential to the larger mission. If the war was about defending South Vietnam and its people, then helping seventy orphans survive was exactly what he should be doing.

When *Stars and Stripes* interviewed him, his explanation was characteristically direct: **"These are still people."**

Not "strategic assets." Not "civilians in the operational area." Not "potential intelligence sources."

**People.**

## MISSION FIRST, PEOPLE ALWAYS

The phrase "mission first, people always" might sound contradictory. How can the mission come first while people always matter? But it's not a contradiction. It's a framework for leadership that acknowledges two simultaneous truths:

**Mission First** means:

- The objective matters
- Results count
- Standards exist for a reason
- Discipline and execution are essential
- Personal preferences don't override organizational needs

**People Always** means:

- How you accomplish the mission affects whether it's sustainable
- The way you treat people in pursuit of goals shapes your culture
- Sacrificing people unnecessarily destroys long-term capability
- Leaders have moral obligations that transcend organizational charts
- Purpose requires caring about more than just metrics

Captain Moye embodied both. He helped his battalion fight effectively in combat operations—at Tet, during Mini-Tet, and in countless patrols and sweeps through hostile territory. **That was mission first.**

But he also took time to find an orphanage, coordinate resources to help seventy children, and demonstrate to his troops and the Vietnamese civilians that American soldiers could be builders as well as fighters. **That was people always.**

The two didn't conflict. They reinforced each other.

## POSITIONAL AUTHORITY VS. MORAL AUTHORITY

There's another dimension to Captain Moye's orphanage work: the difference between positional authority and moral authority.

As a captain in HHC, 2/47th Infantry, Captain Moye had positional authority. He could issue orders within his scope of command. He could requisition supplies through proper channels. He could direct his people to accomplish tasks. But getting base engineers to help dig a well for a Vietnamese orphanage outside the wire? Persuading medical teams to add an orphanage to their rounds? Convincing supply sergeants to "find" extra building materials? Coordinating donated supplies from stateside charity organizations?

None of that came from positional authority. It all came from moral authority. Moral authority is what you earn when people believe:

- You're genuinely trying to do the right thing
- You're not just following rules or covering yourself bureaucratically
- You see them as human beings, not just resources
- You're willing to spend your own capital to help others

Captain Moye built moral authority by finding the orphanage in the first place, by listening to the monk's concerns, by finding creative solutions that respected the monks' dignity ("we don't want your money"), and by following through over months to ensure the project succeeded. That moral authority made everything else possible. People wanted to help because they believed in what he was doing, not because they had to.

**Modern application:** Positional authority gets you compliance. Moral authority gets you commitment. The best leaders use both, but they understand that moral authority is what sustains organizations through difficulties.

When your team needs to work weekends to meet a client deadline, positional authority can mandate it. But moral authority—built by demonstrating that you care about their lives, that you wouldn't ask if it weren't genuinely necessary, and that you'll be there working alongside them—is what determines whether they're grudgingly compliant or genuinely committed.

## PURPOSE THAT OUTLASTS CRISIS

The final lesson from Captain Moye's Vietnam service is about **purpose that outlasts specific missions.**

The orphanage project took three months. Then Captain Moye finished his Vietnam deployment and went home. The war continued for seven more years. The communists eventually won. South Vietnam ceased to exist. American involvement in Vietnam is widely considered a strategic failure.

**So did the orphanage project matter?** By narrow strategic metrics, probably not. Seventy children in one village didn't change the war's outcome. The civic action programs didn't win Vietnamese hearts and minds decisively enough to alter the conflict's trajectory.

But by the metric that actually matters to leadership—**did it give people purpose beyond the immediate mission?**—yes, it absolutely mattered.

The soldiers who helped build that orphanage carried a different story about their Vietnam service than those who experienced only combat operations. They could tell their families, their children, and their grandchildren: "Yes, the war was terrible. Yes, we saw awful things. But we also helped seventy kids survive. We gave them clean water and medicine and buildings. We made their lives better."

**Purpose that includes helping people provides resilience that pure mission focus cannot.** This matters enormously in business contexts. Companies that define their purpose purely in terms of financial metrics—maximize shareholder value, hit quarterly targets, and capture market share—struggle to sustain commitment when those metrics disappoint.

But companies that define purpose in terms of impact on people—the customers whose problems they solve, the communities they support, and the employees whose lives they improve—create reserves of commitment that outlast quarterly setbacks.

Captain Moye's soldiers who worked on the orphanage didn't experience their Vietnam service as pointless, even though the larger war failed. **They had a purpose within the mission that gave meaning to their sacrifice.**

## ADAPTIVE LEADERSHIP IN AMBIGUOUS ENVIRONMENTS

One more dimension of Captain Moye's service deserves examination: his ability to operate effectively in an extraordinarily ambiguous environment.

Vietnam was not World War II, with clear front lines and decisive battles. It was counterinsurgency—an environment where:

- The enemy was difficult to distinguish from civilians
- Tactical victories didn't translate to strategic progress
- The South Vietnamese government was corrupt and often incompetent
- American public support was collapsing
- The operational rules kept changing
- Allied forces (Thai, Australian, and South Korean) brought their own agendas and methods

Captain Moye had to coordinate with:

- His own battalion chain of command
- Task Force Forsyth leadership
- 9th Infantry Division headquarters
- II Field Force staff
- Thai military commanders and the Special Liaison Section
- ARVN (Army of the Republic of Vietnam) units
- Local Vietnamese civilian officials
- Buddhist monks running an orphanage

He had to manage relationships across vast cultural, linguistic, organizational, and operational gaps. He had to accomplish missions when the rules of engagement kept shifting. He had to maintain morale when strategic purpose was increasingly unclear.

**This is adaptive leadership:** the ability to function effectively when traditional command-and-control structures are inadequate, when the situation is ambiguous, when success isn't clearly defined, and when you have to build coalitions across organizational boundaries.

**Modern business application:** Most leadership books assume relatively stable environments where you can develop strategy, communicate it clearly, execute systematically, and measure results objectively.

But that's rarely how business actually works. Markets shift. Technologies disrupt. Competitors do unexpected things. Regulations change. Partnerships evolve. M&A activity reshuffles everything. Leaders who can function only in stable, clear environments are increasingly obsolete. The future belongs to adaptive leaders who can:

- Build relationships across organizational boundaries
- Find purpose in ambiguous situations
- Balance competing priorities (mission and people, results and relationships, short-term and long-term)
- Maintain team cohesion when the larger strategic picture is unclear
- Create meaning for their teams that transcends quarterly metrics

Captain Moye demonstrated this adaptive leadership throughout his Vietnam deployment. He couldn't control the strategic war, couldn't change American political dynamics, and couldn't fix South Vietnam's government. But he could help his battalion execute effectively, coordinate successfully with Thai allies, and improve life for seventy orphans near his base.

**He found the level at which he could make a difference and operated there with excellence.**

## COMING HOME

Captain Moye completed his Vietnam deployment and returned to his family. His son Mike—born during Tet while he was in combat—finally met his father. His older son James, born in 1966, had vague memories of a father who'd been gone.

Captain Moye went on to serve a full twenty-five-year career in the Army, deploying to various assignments around the world, ultimately being promoted to lieutenant colonel and battalion commander. He retired and eventually passed away in 2008—forty years after his Vietnam service, but forever shaped by that experience.

I wasn't born until 1976—eight years after his Vietnam deployment. But growing up, I heard stories—not the dramatic combat stories that movies focus on, but the quieter stories about the orphanage, about working with Thai soldiers, and about trying to do the right thing in an impossible situation.

When I asked him once what he was most proud of from Vietnam, he didn't mention battles or medals. He talked about the orphanage. He talked about showing his soldiers that you could still be human in a war zone, that "these are still people" wasn't a weakness but a strength. **That's the lesson he passed to me, and it's the one I carried into my own service decades later.**

## THE LEADERSHIP FRAMEWORK: MISSION FIRST, PEOPLE ALWAYS

Looking across Captain Jim Moye's Vietnam service, several leadership principles emerge that transcend the specific context of counterinsurgency warfare:

### *1. Purpose Requires More Than Mission Accomplishment*

Captain Moye's battalion had clear missions: Secure Bearcat, respond to threats, and conduct operations in assigned areas. But **purpose**—the "why" that sustains commitment through difficulty—required something more.

The orphanage project gave his soldiers a different story about their service. It connected their daily work to something beyond tactical objectives. It demonstrated that they were in Vietnam to help people, not just fight communists.

**Modern application:** Companies that define purpose purely through business metrics (revenue, market share, and profitability) struggle to sustain commitment when those metrics disappoint. Purpose must connect to impact on people—customers, employees, and communities—in ways that feel meaningful even when quarterly results vary.

When your team hits a rough patch, "We need to increase shareholder value" won't sustain them. But "We're helping families afford life-saving medication" or "We're making small businesses more efficient" or "We're giving people opportunities they wouldn't otherwise have"—those purposes outlast setbacks.

### *2. Moral Authority Amplifies Positional Authority*

Captain Moye could order his direct reports to do things within his command scope. But the orphanage project required voluntary cooperation from people across organizational boundaries: engineers, medical staff, supply sergeants, and even charitable organizations back home.

**He got that cooperation through moral authority—people's belief that he was genuinely trying to do right, not just advance his career or follow regulations.**

**Modern application:** The higher you rise in leadership, the more you need moral authority to complement positional authority. CEOs can mandate strategic changes, but successful execution requires voluntary commitment from people across the organization who don't directly report to them.

You build moral authority by:

- Being transparent about your reasoning, including when you don't have all the answers
- Demonstrating that you see people as humans, not resources
- Making decisions based on principles, not just expediency
- Following through on commitments, especially when it's difficult
- Spending your own capital (time, political currency, resources) to help others

### 3. *Adaptive Leadership Requires Operating Across Boundaries*

Captain Moye coordinated with:

- Multiple U.S. military chains of command
- Thai allied forces with different language, culture, and operational methods
- South Vietnamese civilian and military officials
- Local Vietnamese civilians, including Buddhist religious leaders

He had to build relationships, find common ground, and create shared understanding across vast cultural and organizational gaps—all while conducting combat operations in an ambiguous environment.

**Modern application:** The most valuable leaders in modern organizations are those who can operate across boundaries: between departments, between companies in partnerships, between technical and business functions, and between headquarters and field operations.

Matrix organizations, strategic partnerships, and cross-functional teams all require leaders who can build influence without direct authority, translate between different organizational languages, and create alignment despite competing priorities. The skills aren't taught in most MBA programs, but they're essential. And they're the same skills Captain Moye demonstrated coordinating U.S., Thai, and Vietnamese efforts around Bearcat.

### *4. You Can Accomplish the Mission While Caring for People*

The phrase "mission first, people always" works because the two aren't in conflict when you understand that sustainable mission accomplishment requires caring for people.

Captain Moye didn't choose between combat effectiveness and helping orphans. He did both. His battalion remained operationally ready while simultaneously demonstrating to Vietnamese civilians and his own troops that American soldiers could be builders as well as fighters.

**Modern application:** Leaders often frame choices as binary: "We can either hit our numbers or invest in our people." But the best leaders reject that framing. You can meet quarterly targets while investing in employee development. You can execute a difficult restructuring while treating affected employees with dignity. You can demand high performance while providing support for people's lives outside work.

"Mission first, people always" means:

- Set high standards AND help people meet them
- Demand accountability AND provide resources and support
- Drive for results AND care about the human cost
- Execute tough decisions AND explain the reasoning transparently
- Focus on metrics AND remember that people are more than numbers

Leaders who do this well build organizations that can sustain high performance over time, because people believe their leaders care about more than just quarterly results.

### *5. Find the Level Where You Can Make a Difference*

Captain Moye couldn't end the Vietnam War. He couldn't fix South Vietnam's government. He couldn't change American political dynamics or strategic decisions in Washington. But he could help his battalion execute well. He could coordinate effectively with Thai forces. He could improve conditions for seventy orphans near his base.

**He found the level at which he could make a difference and operated there with excellence, rather than becoming paralyzed by what he couldn't control.**

**Modern application:** All leaders face circumstances beyond their control: economic conditions, competitive dynamics, corporate decisions made above their level, market forces, or regulatory changes.

The leaders who succeed are those who:

- Identify what they CAN control within the larger situation they can't
- Focus their energy at the level where their actions matter
- Execute with excellence in their sphere of influence
- Build capability and relationships that may matter later when circumstances change

This is radically different from either learned helplessness ("nothing I do matters, so why try?") or grandiose delusion ("I can change everything through force of will"). It's realistic assessment of what you can influence, combined with commitment to excellence at that level.

## THE STORY HE TOLD

Years later, when I asked my father what he wanted me to understand about leadership from his Vietnam experience, he told me the orphanage story.

Not the battles. Not the medals. Not the dramatic moments that movies focus on. **The orphanage.**

Because that was the story that exemplified what he believed leadership should be: **accomplishing the mission while caring for people, building moral authority through action, finding purpose beyond the immediate tactical objectives, and making a difference at the level where you can actually influence outcomes.**

He told me about Mr. Nguyen Van Siu, the Buddhist monk who said, "We don't want your money; money brings evil," and forced him to think creatively about how to help. He told me about the well they dug, the buildings they helped construct, and the medical care they provided. He told me about visiting the orphanage months later and seeing children who looked healthier, seeing buildings with solid roofs, and seeing the monks' gratitude.

And he told me: **"These are still people. Always remember that."**

When I flew my own combat missions over Iraq decades later, when I faced my crew's near-mutiny on that search-and-rescue mission in April 2003, and when I had to find words to motivate people to risk everything for strangers in danger, I thought about my father's story.

**Mission first, people always.**

We had a mission: Find that downed helicopter crew. But the reason to risk our lives wasn't abstract strategy or following orders. It was that our fellow servicemembers were in danger, and they were **people**. If we were in their position, we'd want someone to come for us.

Purpose isn't vision. Vision describes where you're going. **Purpose explains why it matters enough to risk everything.**

My father found purpose in an orphanage three thousand meters from his base perimeter in a war that America would eventually lose. That purpose sustained him, gave his service meaning beyond tactical objectives, and taught his soldiers that they could be more than just fighters.

Forty years after Vietnam, he could have been bitter about a war that ended in defeat, about sacrifices that seemed wasted, and about a cause that became politically toxic. Instead, he could point to seventy children who survived because he went looking for them. **That's purpose that outlasts defeat. That's leadership that creates meaning beyond metrics.**

## FROM BEARCAT TO THE BOARDROOM

Captain Moye's Vietnam service might seem distant from modern business challenges, but the leadership dynamics are identical:

**You're operating in ambiguous environments where success isn't clearly defined.** Vietnam didn't have clear front lines or decisive battles. Your market doesn't have clear boundaries or permanent victories either. Competitors shift, technologies disrupt, and customer preferences evolve. Leadership in ambiguity requires the same adaptive skills Captain Moye demonstrated.

**You're coordinating across organizational boundaries with people who don't report to you.** Captain Moye worked with Thai forces, South Vietnamese officials, and multiple U.S. command structures. You're working with strategic partners, cross-functional teams, external vendors, and stakeholders with competing priorities. Both require building influence without authority, finding common ground across different organizational cultures, and creating alignment despite competing agendas.

**You're trying to sustain commitment when the larger strategic picture is uncertain.** By 1968, few people believed America was winning in Vietnam, but soldiers still had to execute daily missions. In business, you often face similar dynamics—the company's being acquired, the industry's consolidating, or the strategic direction keeps changing—but your team still needs to perform. Purpose becomes crucial for sustaining commitment through strategic uncertainty.

**You're balancing short-term demands with long-term capability building.** Captain Moye had immediate combat missions while he simultaneously worked on civic action programs that might not show results for months or years. Every leader faces similar tensions between quarterly pressures and investments in people, culture, and capabilities that pay off slowly.

**You're making ethical decisions in situations with no clear right answers.** Was it appropriate for Captain Moye to divert military resources to a Vietnamese orphanage when American soldiers were dying? There's no objectively correct answer—just judgment calls about what matters and how to allocate finite resources. Business leaders face the same thing constantly: layoffs versus quarterly losses, short-term profits versus long-term sustainability, or shareholder returns versus stakeholder interests.

The contexts differ, but the leadership challenges are structurally identical. And the principles that worked in Vietnam work in business:

- Mission first, people always—drive for results while caring about human impact
- Build moral authority through demonstrated commitment to doing right, not just following rules
- Find purpose beyond metrics that sustains commitment when quarterly results disappoint

- Adapt across boundaries, building relationships and influence beyond your direct authority
- Focus on the level where you can make a difference, executing with excellence rather than becoming paralyzed by what you can't control

## THE WELL THAT OUTLASTED THE WAR

There's one final detail worth noting about Captain Moye's orphanage project.

The 72-foot well they dug by hand in early 1968 probably continued providing clean water long after American forces left Vietnam. Wells can last decades if properly maintained. Even after the communist victory in 1975, even after the American defeat, and even after everything Captain Moye fought for in Vietnam collapsed, **that well likely kept providing clean water for that community.**

Not because of grand strategy or military victory or political success, but because someone cared enough to dig it.

**That's the leadership lesson:** Build things that outlast your tenure, that survive strategic defeats, and that matter beyond your immediate mission. Help people in ways that create value regardless of how the larger story plays out.

When you leave your current role—and you will, eventually, whether through promotion, retirement, or career change—what wells will you leave behind? What did you build that will continue to matter after you're gone?

Not just organizational charts and process documents and strategic plans. Those rarely outlast the leader who created them. **The things that endure are the people you developed, the culture you shaped, the capabilities you built, and the lives you touched.**

Captain James Morris Moye III left Vietnam in 1968. The war continued for seven more years and ended in American defeat. But somewhere near what used to be Camp Bearcat, in a village that's probably been absorbed into Greater Saigon/Ho Chi Minh City's sprawl, there may still be a well providing clean water.

**Mission first, people always.**

# CHAPTER 4

# *The Purpose Each Person Carries*

## IRAQ WAR: NORTHERN IRAQ, 2003

March 26, 2003, 0200 hours. Souda Bay Naval Air Station, Crete.

WE SAT IN THE ready room at Souda Bay, flight suits on, gear staged, and pre-flight briefing complete. Combat Air Crew Four—CAC-4—was about to make history, though we didn't think of it that way at the time. We were about to become the first U.S. Navy P-3 Orion crew to fly an overland Intelligence, Surveillance, and Reconnaissance mission into northern Iraq.

The phone rang. Our operations officer listened, and then hung up. "Stand down. Presidential authority has been pulled."

Forty-eight hours of adrenaline-fueled preparation collapsed into confused exhaustion. We'd been ready to launch—the first P-3 into a combat zone since Vietnam, flying an aircraft never designed for harm's way, and executing a mission we'd trained for but never actually done. And then nothing.

We shuffled back to our quarters. We got maybe three hours of sleep and then woke to another call: "Brief again tonight. You're going."

This time, we launched.

## AN UNLIKELY WARRIOR

The P-3C Orion isn't what people picture when they imagine combat aircraft. It's not an F/A-18 screaming off a carrier deck or a B-52 laying down carpet bombs. It's a four-engine turboprop that looks more like an oversized regional airliner than a weapon of war. The P-3 was designed in the 1960s to hunt Soviet submarines—to fly long patrols over cold oceans, dropping sonobuoys, tracking underwater contacts, and coordinating with surface ships. It's basically a flying computer lab filled with sensor operators, radar technicians, and acoustic analysts. A crew of ten or eleven people would work twelve-hour missions over open water.

But in 2003, with the Soviet Union long gone and unmanned drones still in their infancy, the Navy needed ISR platforms immediately. They needed eyes over Iraq, and the P-3—retrofitted with cameras, electronic surveillance equipment, and targeting systems—was suddenly the only available platform that could fly twelve-hour missions and provide the coverage ground commanders needed.

So they sent us to war in an airplane that had never been designed to be shot at.

We had chaff and flare—defensive countermeasures that might confuse heat-seeking or radar-guided missiles. We had some basic electronic countermeasures. But the fundamental reality was simple: If a surface-to-air missile locked onto us, we were probably going down. The P-3 isn't maneuverable. You can't dodge in a four-engine turboprop. You can't outrun anything in an aircraft with a maximum speed of about 400 knots.

Everyone on the crew understood this. We'd all done the math. We knew the risks. And we went anyway.

## THE FIRST MISSION

Our flight path took us north from Crete, across the Mediterranean, through Turkish airspace (which took diplomatic negotiations at levels far above our pay grade), and into northern Iraq near Mosul.

Mosul was strategically critical—a large city in Iraqi Kurdistan, oil fields, and key transportation routes. In late March 2003, it was still under Iraqi military control: elements of their V Corps and various paramilitary forces. But U.S. forces were inserting: The 173rd Airborne Brigade had just dropped into Bashur Airfield to open a northern front, and special operations forces—Navy SEALs and Marine Recon—were operating throughout the region.

Our job was to provide overwatch: real-time intelligence, target identification, battle damage assessment, communications relay, or whatever the ground forces needed from an eye in the sky.

We checked in with the AWACS—an Air Force E-3 Sentry aircraft that provided airborne air traffic control over the combat zone. We received our specific tasking. We entered Iraqi airspace at 25,000 feet. And immediately, everything became real. We'd trained for this. We'd done hundreds of missions hunting submarines, tracking surface ships, and conducting surveillance. But we'd never done this: flying over a country where people were actively trying to kill us.

The Iraqi Air Force had effectively surrendered—their pilots weren't flying, which is why the Navy felt safe enough to send lumbering P-3s into the fight. But the Iraqis still had plenty of surface-to-air missiles: shoulder-fired SA-7s and SA-14s (the Russian equivalents of American Stinger missiles), plus larger systems such as SA-2s and SA-3s left over from the Soviet era.

Intelligence briefings had shown us the threat rings—the areas where these systems were known or suspected to be positioned. We planned our flight paths to minimize exposure, staying high when possible and using terrain masking when necessary.

But you can't eliminate the risk. You can only manage it and hope.

## "SHIT, SHIT, SHIT, WHERE?"

We'd been on station maybe thirty minutes when it happened.

The observers in the back of the aircraft—enlisted sensor operators stationed at the bubble windows—had one critical additional job: Watch for missile launches. If they saw fire, smoke, or anything that looked like a surface-to-air missile coming up, they were supposed to call it out immediately with precise information: "Missile launch, three o'clock low" or "SAM, nine o'clock, two miles."

Instead, we heard: "Shit! Shit! SHIT!"

Every head snapped around. Every heart rate doubled. The pilots' hands moved instinctively toward defensive maneuvers, even though we had no idea where the threat was.

"WHERE?" someone yelled over the intercom.

"SHIT! SHIT! SHIT!"

"WHERE?!"

It turned out that it wasn't a missile launch. It was ... something else. A fire on the ground, maybe. Or an explosion from ground combat. Or just nerves manifesting as panic over something ambiguous.

But the damage was done. We'd all just experienced a shot of pure adrenaline, the kind that makes your hands shake and your stomach drop and your mind race through every possible worst-case scenario. After we confirmed there was no actual threat, after everyone's heart rates came down, someone keyed the intercom with deadpan delivery: "Outstanding tactical communication there. Really helped us identify the threat."

Nervous laughter rippled through the crew.

And just like that, "Shit, shit, shit, where?" became our crew's running joke. Any time someone saw anything remotely concerning for the rest of the deployment: "Shit, shit, shit." And someone else would immediately respond: "WHERE?" And we'd all laugh—the kind of gallows humor that gets you through impossible stress.

## THE GRIND

That first mission was just the beginning. Between March 26 and May 1, 2003—when President Bush declared "mission accomplished" aboard the USS *Abraham Lincoln*—we flew approximately twenty combat missions into northern Iraq.

Each mission was roughly the same: twelve-hour flights airborne, three hours transit each way, six hours on station over hostile territory, and refueling in Cyprus or Saudi Arabia or wherever we could just to get back. We'd support ground operations, track targets, provide communications relay, and conduct battle damage assessment after airstrikes. Then we'd fly home to Crete, debrief for two hours, sleep for maybe six, and do it again.

The operational tempo was punishing. We were flying far more than we were supposed to—the normal crew rest requirements were waived because they needed the coverage. Everyone was exhausted due to the stress of flying combat missions in an aircraft that couldn't defend itself compounded with the physical exhaustion of twelve-hour flights and minimal rest. And this is where the leadership challenge emerged—the same one my father had faced in Vietnam, the same one John E. Moye had faced during the Siege of Vicksburg, and the same one Captain

George Moye had faced during the long middle years of the Revolutionary War.

How do you keep people motivated when they're exhausted, when the risk is real, when the purpose that seemed clear at the beginning starts feeling abstract? I was twenty-six years old. I was the tactical coordinator—the senior naval flight officer on the crew, responsible for managing our mission execution. I'd been flying for about five years. I'd been through training, I'd earned my position, but I'd never faced anything like this.

And my crew was losing faith—not in me, specifically; not in our capabilities; but in the broader purpose. Why were we here? What were we actually accomplishing? Was this mission worth the risk? These were smart people—the best operators from their training classes, selected for this crew specifically because they were the top performers. But even the best people reach their limits.

## THE SEARCH AND RESCUE MISSION

I've already described what happened on April 6, 2003—the call that came in about the downed army helicopter near Baghdad, my crew's initial refusal to take the mission, and the ten minutes where I had to remind everyone why we were there.

But what I haven't explained is what made that moment possible. It would have been easy for that crew to simply refuse, saying, "No, we're not doing this. We've already done more than our share. Someone else can take this one."

I couldn't have forced them. Not really. I had positional authority—I was the TACCO, the senior tactical officer on the aircraft. But in that moment, positional authority meant nothing. If my crew had genuinely refused that mission, there wasn't much I could have done about it mid-flight.

What made them choose to accept the mission wasn't my authority. It was purpose. But here's what I've realized in the twenty years since: It wasn't one purpose. It was multiple purposes, each personal and each valid.

## THE PURPOSES WE EACH CARRIED

When I think back to that crew and what motivated each person to accept that search-and-rescue mission despite their fear, I realize now that everyone had different reasons.

The pilot-in-command—a lieutenant also, a year or two older than I was, but more experienced. His purpose was partly professional pride: He was a naval aviator, and naval aviators don't refuse missions when American servicemembers need help. But it was also personal: He had kids at home, and he wanted to be the kind of father who could tell his children he'd done the right thing when it mattered. He had a legacy of naval service as well—his dad was a well-known admiral in the fleet.

The copilot—called the 3P—brand new to the squadron, maybe twenty-three years old—was terrified. But he was also ambitious. His purpose was proving himself, showing he belonged on this crew, and earning respect from the more experienced people around him. Going into a dangerous situation wasn't what he wanted, but backing down would have been worse for how he saw himself.

The senior chief flight engineer—grizzled, 25-years-in-the-navy kind of guy—told me later that he agreed to the mission because he kept thinking: "What if that was me down there? What if I needed help and someone refused because it was too risky?" His purpose was empathy, putting himself in someone else's position.

The sensor operators and tactical systems operators—the enlisted guys who ran the radar, the cameras, and the electronic surveillance equipment—had varying motivations. Some were committed to the mission because they believed in supporting ground forces. Some were motivated by loyalty to the crew—they didn't want to let their teammates down. Some were frankly just following orders and hoping we'd all get home alive.

What I'd done when I reminded them of their oaths to the Constitution, when I said, "Our brothers and sisters are on the ground and need our help," was connect their individual purposes to a shared mission.

I didn't create their purposes. I couldn't have. Purpose comes from within. What I did was remind them of what they already believed and show them how this immediate decision connected to it.

## TWENTY YEARS LATER: THE CEO'S LESSON

Fast-forward to 2018. I'm the CEO of Paige AI, a healthcare technology company building the first FDA-approved artificial intelligence algorithm for detecting prostate cancer in pathology slides.

Paige was born from research at Memorial Sloan Kettering Cancer Center. Our technology had demonstrated in rigorous clinical studies that it could reduce diagnostic errors by up to 70 percent. Pathologists looking at tissue samples—trying to determine if cancer is present, what grade it is, and whether it's spreading—make mistakes. They're human. They get tired. They miss things.

Our AI didn't get tired. It could review thousands of slides with consistent accuracy. And most importantly, when pathologists used our AI as a second opinion, their accuracy improved dramatically. This was visionary purpose at its finest: using artificial intelligence to save lives by improving cancer diagnosis. Every person at Paige could articulate it. When we raised money from investors, when we recruited talent, and when we talked to healthcare systems about adopting our technology, our purpose was crystal clear.

But here's what I learned as CEO: Organizational purpose isn't enough. You need to understand individual purpose.

One of the first things I did when I became CEO was conduct what I called a "listening tour." I scheduled thirty-minute one-on-one meetings with every single person in the company—about 150 people at the time. including engineers, data scientists, business development people, operations staff, and administrative assistants.

I asked each person the same basic questions:

- Why did you join Paige?
- What gets you excited about coming to work?
- What matters to you personally?
- What are you trying to accomplish in your career?
- What drives you?

The answers were illuminating. Many people were deeply motivated by our cancer mission. They'd had family members die of cancer. They'd watched loved ones go through misdiagnoses or delayed diagnoses. They were at Paige because they genuinely wanted to save lives and improve healthcare. For them, our organizational purpose and their individual purpose were perfectly aligned.

My executive assistant—incredibly talented, holding two master's degrees including one from Columbia—was working in an administrative role that probably didn't fully use her capabilities. When I asked why she was at Paige, she said: "My mother died of ovarian cancer. I wanted to work somewhere that was actually fighting cancer, not just talking about it."

That's deep, personal purpose. I couldn't have created that or manufactured it. It was already there. My job was to recognize it and ensure her work connected to it. But not everyone was motivated by cancer.

I had engineers who were fascinated by the technical challenge of building AI systems that could analyze gigapixel medical images. They were excited about being on the cutting edge of machine learning, about solving problems that were genuinely novel and difficult. Their purpose wasn't saving lives per se—it was pushing the boundaries of what technology could do. And that was completely legitimate.

I had business development people who were motivated by building relationships, by convincing skeptical healthcare systems to try something new, and by the challenge of market entry and adoption. Their purpose was about proving a new category of medical technology could work in the real world.

I had data scientists who'd come from big tech companies—Google, Facebook—and were at Paige because they wanted their work to matter more. They were tired of optimizing ad click-through rates. Their purpose was doing work that felt meaningful, even if they weren't personally connected to cancer.

And yes, I had people who were there because it was a good job with competitive pay, interesting colleagues, and career growth opportunities. They wanted to do good work, provide for their families, and develop their skills. Their purpose was professional and financial—and that was okay too.

## THE MISTAKE OF ONE-SIZE-FITS-ALL PURPOSE

Here's the leadership mistake I see constantly: Leaders assume that if they articulate a compelling organizational purpose, everyone will connect to it the same way. So they give inspirational speeches about saving lives, or disrupting industries, or changing the world. They create mission statements and values documents. They talk about vision and purpose and impact.

And it works for some people. The ones whose individual purpose naturally aligns with organizational purpose get fired up and motivated. But for everyone else—the people whose purposes are different, more personal, or less grandiose—those big inspirational speeches can actually be demotivating. Because there's an implicit message: If you're not deeply moved by our world-changing mission, you don't really belong here.

This is wrong, and it's counterproductive. The best teams have people with diverse purposes, all pulling in roughly the same direction.

At Paige, I needed:

- True believers who were passionately committed to fighting cancer
- Technical perfectionists who were obsessed with building the best AI systems possible
- Business operators who got excited about market strategy and adoption curves
- People who just wanted to do excellent work in a professional environment

All of those purposes were valuable. All of them contributed to our success. My job as CEO wasn't to convert everyone to the same purpose. My job was to understand each person's purpose and help them see how their work at Paige connected to it.

I started thinking of this as "purpose translation"—taking our organizational purpose and translating it into terms that resonated with each individual's personal purpose.

For the engineer excited about technical challenges: "The reason this problem is so hard is that pathology images are gigapixels—much larger than normal computer vision datasets. You're solving problems that Google and Facebook haven't solved. This is legitimately cutting-edge."

For the business development person motivated by market strategy: "We're creating an entirely new category. Healthcare systems don't have a box to put 'AI diagnostic tools' in yet. You're literally defining how this market will work. That's rare in a career."

For the data scientist who left big tech: "Every model you improve means pathologists make better diagnoses. You can draw a direct line from your code to patient outcomes. When's the last time you could do that?"

For the person focused on career growth and financial stability: "Being early at a company like this, if we succeed, has massive career upside. You'll have been part of building something from scratch in a space that's exploding. And even if we don't succeed, you're developing skills in AI and healthcare that make you incredibly valuable."

None of these translations were manipulative or dishonest. They were all true. Our organizational purpose—improving cancer diagnosis with AI—was legitimately meaningful. But meaning is personal, and different people find meaning in different aspects of the same work.

## ALL-HANDS AND INDIVIDUAL STORIES

Every week at Paige, we held an all-hands meeting—the entire company, thirty minutes, standing room only in our conference space and on Zoom.

Most of the meeting was operational: updates on clinical trials, product development progress, business development wins, and hiring plans. It was the standard CEO communication. But once a month, I'd ask someone to share a personal story about why he or she was at Paige. Sometimes it was someone with a personal cancer story—a parent, a sibling, or a friend who'd been misdiagnosed or diagnosed too late. Those stories were powerful and emotional. They reminded everyone why our work mattered beyond just building a successful company.

But sometimes it was someone talking about the technical challenge, or the career opportunity, or what it felt like to be part of building something new. Those stories were equally valuable, because they validated that you didn't need to have a personal tragedy to justify being at Paige. Professional purpose was legitimate too.

The goal was to show that we're all here for different reasons, and that's a strength, not a weakness.

## WHEN PURPOSE ISN'T ENOUGH

I'd love to say that understanding individual purpose solved all leadership challenges at Paige. It didn't.

We went through layoffs—multiple rounds—as we tried to extend our cash runway between fundraising rounds. We lost key people. We had product setbacks. We had FDA delays. We had healthcare systems that committed to pilots and then backed out.

Some people left because they lost faith in the company's ability to succeed. Some left because their individual purposes changed—they needed more financial stability than a startup could provide, or they wanted different technical challenges, or they were recruited away by bigger companies with more resources.

Purpose sustains people through difficulty, but it doesn't make a difficulty disappear. The question is: Does understanding individual purpose help when times are hard?

I think yes, but not in the way you might expect. Understanding individual purpose didn't prevent people from leaving. But it did help me have honest conversations about whether Paige was still the right fit for them. When someone came to me considering an offer from a big tech company with twice the salary, I didn't try to guilt them about abandoning our mission. I asked: "What's driving this decision? Is it the money? The stability? The scope of impact? The technical resources?"

If it was money and stability—if they had a family to support and needed financial security—I could genuinely wish them well. Their purpose had changed, and Paige couldn't meet it anymore. No hard feelings.

If it was technical resources—if they were frustrated by the constraints of startup budgets—I could be honest about what we could and couldn't provide. Sometimes we'd find creative solutions. Sometimes they'd leave anyway. But the conversation was honest.

If it was scope of impact—if they wanted to work on systems with hundreds of millions of users instead of thousands of pathologists—I could respect that. Their purpose was valid, just different from what Paige offered. Understanding individual purpose lets you have humane conversations about fit instead of binary conversations about loyalty.

## BACK TO BAGHDAD

The reason I connected my Iraq experience to my Paige experience is this: The fundamental leadership challenge is the same.

On April 6, 2003, over Baghdad, I needed my crew to accept a dangerous mission they didn't want to fly. I couldn't force them. I could only help them connect the mission to purposes they already held.

As CEO of Paige, I needed people to work incredibly hard—startup hours, startup uncertainty, and startup stress—for compensation that was less than they could get at big established companies. I couldn't force them. I could only help them see how their work connected to purposes they already held.

Purpose is what fills the gap between "what I'm asking you to do" and "what you're willing to do." But purpose isn't one thing. It's not just organizational mission. It's not just personal values. It's the intersection of:

- What this organization is trying to accomplish
- What I personally care about
- How my work connects those two things
- Whether that connection is worth the cost I'm being asked to pay

Different people will answer those questions differently. And that's okay. The leader's job isn't to give everyone the same purpose. It's to understand each person's purpose and help them see how it connects to shared goals.

## THE SHALLOW END OF THE PURPOSE POOL

I said earlier that there's a "shallow end" of the purpose pool, and it's okay to be there. Some people read that and hear judgment: Shallow purposes are less valid than deep purposes. That's not what I mean.

What I mean is that not everyone's purpose needs to be world-changing. Wanting to do good work, to provide for your family, to develop your skills, and to be part of a functional team are completely legitimate purposes.

The soldier on my crew who wasn't personally invested in the geopolitical outcome of the Iraq War but wanted to do his job well and go home safely—his purpose was valid. The engineer at Paige who wasn't personally affected by cancer but found the technical challenges fascinating—her purpose was valid. The individuals at your company who aren't passionate about your industry but want to build stable careers and provide for their kids—their purpose is valid.

Leaders make a mistake when they demand that everyone be deeply and emotionally committed to the organizational mission. Some people will be. Many won't. And trying to force that kind of commitment creates resentment and disengagement. Better to recognize: "This person's purpose is professional rather than personal, and that's fine. How do I help them excel in their role and connect their professional growth to our organizational goals?"

## THE FRAMEWORK: PURPOSE AT MULTIPLE LEVELS

Looking back across four generations of Moye family military service and my own career in business, I've come to think about purpose at three distinct levels:

### *Organizational Purpose*

This is the big "why" for the entire organization:

- "We're defending American independence" (Revolutionary War)
- "We're preserving our way of life" (Civil War—even though that way of life was morally wrong)
- "We're supporting the South Vietnamese people" (Vietnam)
- "We're stopping terrorism and weapons of mass destruction" (Iraq)
- "We're improving cancer diagnosis with AI" (Paige)

Organizational purpose needs to be:

- Clear enough that anyone can articulate it
- Meaningful enough that it justifies the organization's existence

- Connected to outcomes that matter beyond just organizational survival

### *Team Purpose*

This is the "why" for your specific group within the larger organization:

- "Our squadron provides ISR coverage that ground forces depend on"
- "Our battalion protects Bearcat and coordinates with allied forces"
- "Our engineering team builds the AI models that detect cancer"

Team purpose needs to be:

- Specific to what your team actually does
- Connected to the organizational purpose but more tangible
- Something team members feel ownership over

### *Individual Purpose*

This is the personal "why" that drives each person:

- "I want to serve my country and prove myself"
- "I want to provide for my family"
- "I want to develop skills that advance my career"
- "I want to solve technical challenges that few people can solve"
- "I want to make a difference in healthcare because cancer affected my family"

Individual purpose needs to be:

- Respected even when it's not what you'd prefer
- Understood by their direct manager at minimum
- Connected to their actual work, not just platitudes

Great leaders operate at all three levels simultaneously:

They articulate clear organizational purpose. They help teams understand their specific role in that purpose. And they understand what drives each individual, helping them see how their work connects to their personal purpose.

## APPLYING THIS IN PRACTICE

Okay, so you're leading a team—maybe five people, maybe five hundred. How do you actually apply this?

**Step 1: Understand organizational purpose.** If you're not the CEO, you may not control organizational purpose. But you need to understand it and be able to articulate it clearly. If you can't explain why your organization exists in two sentences, you have a problem.

**Step 2: Translate organizational purpose to team purpose.** What does your team specifically do that contributes to organizational purpose? This should be concrete and measurable. "Our sales team helps hospitals adopt technology that improves cancer diagnosis" is better than "We're part of fighting cancer."

**Step 3: Understand individual purposes.** This requires actual conversations. You cannot assume you know what drives someone. Ask questions:

- Why did you take this job?
- What gets you excited about your work?
- What do you want to accomplish in your career?
- What matters to you personally?
- How do you define success for yourself?

Listen to the actual answers, not the answers you want to hear.

**Step 4: Make connections explicit.** Help each person see how their work connects to their purpose. This isn't manipulation—it's clarity. If someone's purpose is skill development and career growth, show them how their projects build marketable expertise. If someone's purpose is stability and work-life balance, show them how their excellent execution contributes to organizational health that protects their job.

**Step 5: Respect when purposes diverge.** Sometimes someone's individual purpose can't be met by your organization anymore. When that happens, have an honest conversation about fit. Help them transition if necessary. Don't try to guilt them into staying.

**Step 6: Build purpose diversity into your team.** Don't hire only true believers. A team in which everyone has the exact same purpose is fragile—it can't handle different challenges or adapt to changing circumstances. Build teams with diverse purposes that all contribute to shared goals.

## THE SEARCH AND RESCUE MISSION, REVISITED

Let me return to where we started: April 6, 2003, over Baghdad, when an Army helicopter had been shot down and my crew initially refused the mission. What I did in that moment was activate multiple purposes simultaneously.

**Organizational purpose:** "We're here to support American forces in Iraq"

**Team purpose:** "Our squadron provides critical ISR that ground forces depend on—we're the only asset in position to help"

**Individual purposes:**

- For those driven by service to country: "This is exactly why we joined the military"

- For those driven by loyalty to fellow servicemembers: "Those are our brothers and sisters who need help"
- For those driven by professional excellence: "This is the mission that matters most"
- For those driven by proving themselves: "This is your chance to show what you're made of"
- For those driven by getting home safely: "The fastest way to go home with honor is to do our job"

I didn't give them one purpose. I reminded them of the purposes they each already had, and showed them how this immediate decision connected to all of them.

That's why it worked. Not because I was particularly eloquent or inspiring in that moment—if you'd listened to the actual words, they were pretty basic. But because I understood that everyone needed to hear something different, and the mission I was asking them to accept could genuinely connect to multiple valid purposes.

## PURPOSE ISN'T FIXED

One final insight: Individual purpose changes over time.

The twenty-two-year-old pilot on my crew in 2003 was motivated by proving himself and serving his country. When I talked to him fifteen years later, he had three kids and was flying for a commercial airline. His purpose had shifted: stable income, predictable schedule, and being present for his family. He was the same person, but his life circumstances and priorities had changed.

The engineer at Paige who joined because the technical challenges were fascinating might have kids a few years later and suddenly prioritize financial stability over interesting problems. Her purpose evolved.

Great leaders recognize this and adapt. The conversation you have with someone about why they're here needs to be ongoing, not one-time.

When someone's engagement drops, when their performance slips, or when they seem disconnected—before you assume they're not committed or not capable, ask: "Has your purpose changed?"

Maybe their sick parent needs full-time care, and work-life balance suddenly trumps everything else. Maybe they've developed a new passion, and your organization can't provide it. Maybe they've achieved what they came to achieve and need a new challenge.

Understanding this lets you respond appropriately:

- Adjust their role if possible
- Help them transition if necessary
- Give them time if it's temporary
- Find new ways to connect their evolved purpose to organizational needs

## THE LEADERSHIP LESSON

My great-great-great-great-great-great-grandfather George Moye II rallied Pitt County militia by connecting revolution to "the right of Pitt to govern Pitt"—making abstract independence concrete and personal.

My great-great-great-grandfather John E. Moye kept fighting through catastrophic defeats at Champion's Hill and Vicksburg because his purpose was rooted in protecting his community and family, not grand Confederate strategy.

My father Jim Moye helped rebuild a Buddhist orphanage in the middle of a war zone because his purpose transcended military objectives—"these are still people."

And I convinced my crew to fly a dangerous search-and-rescue mission by reminding them of the purposes they each already held.

Different eras. Different circumstances. Same fundamental insight.Purpose drives commitment, but purpose is personal. The leader's job isn't to impose purpose—it's to understand the purposes people already have and help them connect those purposes to shared work.

This applies in combat, where the stakes are life and death. It applies in startups, where the stakes are careers and livelihoods. It applies in established companies, where the stakes are relevance and growth. Wherever you lead, whatever you're trying to accomplish, you're asking people to commit their time, their energy, their creativity, and their effort. They'll truly commit only if they see how what you're asking connects to what they personally care about.

You can't give them purpose. But you can help them find it in the work you're doing together.

And when you do that—when you understand that the pilot motivated by professional pride, the engineer motivated by technical challenge, the administrator motivated by fighting cancer, and the analyst motivated by career growth can all pull together toward the same goal—you build something remarkable. Not because everyone shares the same purpose, but because everyone sees how their different purposes connect to shared success.

## EPILOGUE: TWENTY YEARS LATER

In 2023, twenty years after that mission over Baghdad, I attended a reunion of our old squadron. Most of the crew from CAC-4 was there. We told stories. We laughed about "Shit, shit, shit, where?" and the hundred other moments of absurdity and terror that defined those months. We raised glasses to the people who weren't there, who'd died in later deployments or simply moved on with their lives.

At one point, someone asked: "Why did we really do it? Why did we keep flying those missions when we were so scared?"

The answers varied:

- "I didn't want to let you guys down"
- "It felt important at the time"
- "I was young and stupid and didn't think I could actually die"
- "I believed in what we were doing"
- "I wanted to prove I could"
- "Honestly, I'm still not sure"

All of those answers were true. All of them were valid. We were the same crew, on the same missions, facing the same dangers. But we each had different reasons for being there.

That's not a weakness. That's how teams actually work.

Leaders who understand that—who respect that different people find meaning in different things and who help each person connect their individual purpose to shared goals—build teams that can accomplish the impossible. Even in a war zone at 25,000 feet in an aircraft never meant to be shot at, carrying out a mission that scared us all.

Especially then.

# *Section II: Adaptive Intelligence*

# CHAPTER 5

## *The Calculated Retreat*

March 15, 1781. Guilford Courthouse, North Carolina.

LIEUTENANT GEORGE MOYE III stood in the second line of North Carolina militia, watching the British advance emerge from the tree line four hundred yards away. He was twenty-nine years old now, no longer the young private who'd signed the Pitt Association with his father six years earlier. The scar on his left arm from earlier skirmishes had healed poorly, leaving the limb slightly stiff. But he could still load and fire a musket, and more importantly, he could keep the men around him steady.

Which was exactly what General Nathanael Greene needed today.

Greene's orders had been explicit and unusual: two measured volleys, then a third devastating volley, then retreat. Not "hold until overwhelmed." Not "fight to the last man." A pre-declared, calculated retreat built into the battle plan from the beginning.

To the militia farmers standing in line—men who'd been taught that retreat meant cowardice, that holding ground defined honor—this strategy felt wrong. Counterintuitive. Almost dishonorable.

But George III understood something his father had taught him during the long middle years of the revolution: Sometimes survival is victory. Sometimes losing the field wins the war. Sometimes the most intelligent adaptation is knowing when to pull back and preserve your force.

As the British grenadiers' red coats grew larger, as drums beat cadence and sergeants barked orders, George gripped his musket and waited for the command. Around him, men shifted nervously. Some muttered prayers. Some checked their flints for the third time.

And George thought: *We're about to demonstrate that the smartest decision is often the one that looks like failure.*

## THE QUARTERMASTER'S GENIUS

To understand what happened at Guilford Courthouse, you need to understand Nathanael Greene—one of the most brilliant and least celebrated generals of the Revolutionary War.

Greene didn't start as a combat commander. George Washington made him the Continental Army's quartermaster general in 1778, essentially the logistics chief responsible for keeping the army fed, equipped, and mobile. It wasn't a glamorous role—quartermasters dealt with wagons and provisions and supply lines, not battlefield glory.

But Greene was a genius at it. He understood something that most military officers missed: Wars aren't won by tactics alone. They're won by logistics, sustainability, and the ability to keep fighting when conventional wisdom says you should quit.

When Washington needed a general to take command in the Southern theater after the catastrophic defeat at Camden (the same battle where Captain George Moye II's wagons had been detached to guard the rear), he chose Greene precisely because of his quartermaster experience.

Greene arrived in the South in December 1780 to find a shattered, demoralized army. Horatio Gates had lost badly at Camden. The British controlled most of Georgia and South Carolina. Cornwallis was systematically crushing Patriot resistance.

Conventional military thinking said Greene should consolidate his forces, fall back to defensive positions, and wait for reinforcements.

Greene did the opposite. He split his already small army—a move that horrified his officers—and began a campaign of maneuver, harassment, and calculated engagements designed not to win battles but to wear Cornwallis down.

**This is adaptive intelligence in its purest form: recognizing that the playbook everyone expects you to follow won't work, and having the courage to write a new one.**

By March 1781, Greene had drawn Cornwallis inland, away from British supply lines, forcing the British general to chase him through the Carolina countryside. Cornwallis was winning tactical engagements but losing strategic position—his army was getting smaller with each battle, farther from resupply, and more exhausted.

Greene was ready to give battle, but on his terms. At Guilford Courthouse, he would demonstrate that losing the field could win the war.

## THE THREE-LINE DEFENSE

Greene positioned his forces in three lines, each about four hundred yards behind the previous one:

**First line:** North Carolina militia—farmers and part-time soldiers, least experienced, and most likely to break under pressure. Greene didn't expect them to hold. He expected them to fire two volleys and fall back. Their job wasn't to stop the British; it was to bleed them.

**Second line:** Virginia militia—more experienced and better disciplined. This was where Lieutenant George Moye III stood with the Pitt County men. Greene expected them to deliver punishing fire and then conduct an organized retreat. Not a rout—a planned withdrawal.

**Third line:** Continental regulars—professional soldiers, the core of Greene's army. These men would fight conventionally, holding ground and delivering the decisive engagement.

The strategy was elegant: Force the British to advance through three separate defensive positions, each one taking casualties, and each one requiring them to reform and attack again. By the time Cornwallis's forces reached the third line, they'd be exhausted and depleted. But this required something extraordinary from the militia: the discipline to retreat on command, not in panic; the ability to distinguish between "falling back according to plan" and "breaking in disorder." Most generals didn't trust militia with that kind of complexity. Greene did, but only because he'd prepared them.

George III had spent the previous evening walking among the Pitt County men, explaining the plan: "Two volleys measured, a third to break them—then we fall back. That's the plan. Falling back isn't failing. It's the mission."

His father, Captain George Moye II, now in his late fifties and struggling with gout, had marched anyway. He walked the line tapping shoulders with his maple cane, repeating the same message: "Remember Charleston. Remember Camden. This time, we control the field. We decide when to stand and when to move."

Purpose beyond the immediate moment. Adaptive intelligence that trusted soldiers to execute a complex plan under pressure.

## THE BRITISH ADVANCE

At approximately 1:30 PM, Cornwallis's forces—about 1,900 elite British and German troops—emerged from the woods and began their advance. The first line of North Carolina militia waited. Four hundred yards became three hundred, then two hundred. Officers rode behind the line, reminding men: "Wait for the command. Wait."

At about seventy-five yards, the command came: "Present! Fire!"

The first volley tore through the British ranks. Men fell. The line wavered but kept advancing. The North Carolina militia reloaded—a practiced drill they'd performed hundreds of times, but never under this kind of pressure. The British kept coming, closing to fifty yards.

"Fire!"

The second volley hit even harder at closer range. More British soldiers went down. But now the redcoats were close enough that the militia could see their faces and could see bayonets gleaming. Some of the North Carolina men broke immediately—dropping their muskets and running. This was expected. But many others held just long enough to reload and deliver that crucial third volley before retreating in reasonable order.

The British kept advancing, now toward the second line. Lieutenant George Moye III watched them come. Beside him, a young private—maybe nineteen years old, hands shaking—muttered "Jesus, Jesus, Jesus."

George put a hand on the boy's shoulder. "Steady. Two volleys, then the third. That's all we need."

The British came on, climbing over fallen logs, pushing through brush, and maintaining formation despite the casualties they'd taken. At seventy yards, the Virginia militia fired their first volley. At fifty yards, their second. And then—with British bayonets glittering in the afternoon sun, with every instinct screaming to run *right now*—they held fire just long enough to reload.

George heard his father's voice carrying across the line, that same maple cane tapping: "Purpose, boys. Purpose holds."

The third volley erupted at thirty yards—close enough that the Virginia militia could see the impact, could see Cornwallis's grenadiers stagger and fall, and could see the British line actually pause. Then, as planned, they fell back. Not in panic. Not in rout. In organized sections, they were moving to predetermined positions and maintaining some cohesion, even as they gave up the field.

George III felt something hot graze his left arm—a musket ball that cut through his coat and left a bloody scratch but didn't shatter bone. He caught a private next to him who was starting to bolt. "Purpose, lad—purpose!" They fell back together, toward the third line where Continental regulars waited.

## THE PRICE OF VICTORY

The Battle of Guilford Courthouse lasted about two hours. When it ended, Cornwallis held the field—technically a British victory.

But here's what that "victory" cost:

- British casualties: approximately 532 killed and wounded out of 1,900 engaged (roughly 28 percent of Cornwallis's force)
- American casualties: approximately 264 killed and wounded out of 4,400 engaged (roughly 6 percent)

Cornwallis had won the battle but lost a quarter of his army. Greene had lost the field but preserved his force largely intact. Charles James Fox, a British parliamentarian, would later say: "Another such victory would ruin the British army."

He was right. Cornwallis, bleeding strength he couldn't replace, would retreat to Wilmington and then march to Virginia, seeking reinforcements. By October 1781, he'd be trapped at Yorktown and forced to surrender—effectively ending the war.

Greene's "defeat" at Guilford Courthouse had won the Southern campaign.

**This is what adaptive intelligence looks like: Success that doesn't resemble victory by conventional metrics.**

## THE STARTUP PARALLEL: WINNING BY LOSING

Fast-forward 230 years. Silicon Valley, 2011.

A startup called Burbn had been building a location-based check-in app—essentially trying to compete with Foursquare. They'd raised some seed funding. They'd built a product. They'd launched.

And it wasn't working. User growth was slow. Engagement was poor. The app was too complicated, trying to do too many things. It was a classic startup struggling to find product-market fit. The founders—Kevin Systrom and Mike Krieger—faced the same choice George Moye III had faced at Guilford Courthouse: Keep fighting a battle you're losing, or execute a strategic retreat and redeploy.

They looked at their usage data and noticed something: People weren't really using the check-in features. But they were using the photo filters—the ability to take a picture, apply a vintage filter, and share it. That feature was actually working.

So they did something that felt like failure: They killed Burbn. They threw away months of work. They abandoned features they'd sweated over. They essentially admitted defeat. And they rebuilt, focusing exclusively on photo sharing with filters. Simple, fast, mobile-first. They called it Instagram.

Within two months of launching Instagram in October 2010, they had one million users. Within eighteen months, Facebook bought them for $1 billion.

**They won by losing—by giving up the battle they couldn't win to win the war that mattered.**

This is the Guilford Courthouse strategy applied to startups: pre-declare your retreat conditions, execute them when necessary, preserve your resources (in this case, time, money, and team morale), and redeploy to fight where you have advantage.

## THE LOUDCLOUD LESSON

Ben Horowitz executed the same strategic retreat a decade earlier, under even more brutal conditions. He'd founded

Loudcloud during the dot-com boom to provide managed hosting for internet businesses. The business model looked brilliant in 1999. Then the market collapsed in 2000–2001. Loudcloud's customers failed. Capital dried up. The company teetered on bankruptcy.

Horowitz faced his Guilford Courthouse moment.

He could keep fighting for the hosting business, trying to salvage a model that the market had already killed. Or he could make a brutally honest assessment of what was actually working and retreat to defensible ground.

What was working? The internal software Loudcloud had built to manage complex data-center operations. That tool was more valuable than the hosting business itself.

So Horowitz executed one of Silicon Valley's most disciplined pivots. He sold the hosting operations—the original business, the thing they'd raised money to build. He laid off employees. He completely rewrote the company's story. And he re-founded the company as Opsware, a pure enterprise software company focused on automating IT infrastructure.

This wasn't just insight. This was an execution under pressure.

Horowitz had taken Loudcloud public, which meant managing Wall Street expectations during the pivot. He chose uncommon candor—explaining the retreat clearly rather than hiding from it. He shifted from a service-heavy, low-margin business to a scalable software product. He made the hard, clarity-driven decisions that survival demanded. Over time, Opsware proved the bet correct. It grew into a category-defining company. Hewlett-Packard acquired it for roughly $1.6 billion.

The episode cemented Horowitz's reputation for what he later called "wartime CEO" leadership—not my favorite phrase, but the concept holds. When survival is on the line, you need the discipline to separate ego from evidence and reallocate resources decisively.

Years later, as a venture capitalist at Andreessen Horowitz, Horowitz recognized the same pattern in a struggling game startup. The game wasn't working. But the internal communication tool the team had built to coordinate development? That was working. Users loved it. The founder, Stewart Butterfield, faced the familiar choice: double down on the weak core product or recognize that the side project was the real breakthrough.

Horowitz encouraged the team to focus on what users already loved—the communication tool—and to have the courage to abandon the original vision. They did. They killed the game. They rebuilt around the communication tool. They called it Slack.

Horowitz didn't view these pivots as failures. He saw them as acts of intellectual honesty: listening to the market, accepting reality, and moving decisively. The best pivots aren't reactive flailing. They're deliberate moves grounded in a clear understanding of where real value is being created.

Sometimes the smartest decision looks like losing. Sometimes retreat is how you win the war.

## BEING IN LOVE WITH THE PROBLEM, NOT THE SOLUTION

Here's where most founders and leaders get it wrong: They fall in love with their solution instead of their problem. Burbn's founders could have said: "We've built this location-based check-in app. We've raised money for it. We've told investors this is what we're building. We need to make *this specific thing* work."

That's being in love with the solution.

Instead, they stayed focused on the problem: *How do we help people share moments from their lives in ways that feel natural and engaging?* When it became clear that check-ins weren't the answer, they were free to pivot to photos because they weren't attached to the solution—they were attached to the problem.

**This is Nathanael Greene's quartermaster mindset applied to entrepreneurship:** The goal isn't to win this specific battle; it's to solve the strategic problem. Stay flexible on tactics while remaining committed to the mission.

George Moye III wasn't trying to hold a specific piece of ground at Guilford Courthouse. He was trying to weaken Cornwallis's army enough to make the British Southern campaign unsustainable. When Greene ordered the retreat, George executed it because he understood the larger problem they were solving.

But here's the challenge every leader faces: How do you know when to pivot versus when to stay the course? This is one of the oldest questions in leadership. Which way should we go? When do we adapt, and when do we hold firm? Because persistence matters. Grit matters. Many great companies nearly failed multiple times before breaking through. If you pivot at the first sign of difficulty, you never build anything meaningful. But stubbornness kills, as in refusing to adapt when the evidence is clear. That's not courage, that's delusion.

So how do you tell the difference? Across all the startup pivot stories—Instagram, Slack, Opsware, and hundreds of others—clear signals emerge.

## THE PERSISTENCE SIGNALS

Persist when the core value hypothesis is still being validated—when customers are actually using the product; when they're deriving value from it; or when they're pulling it forward, even if growth is slower than you'd like or execution is messy. Persist when users are genuinely engaged—when retention is improving, even incrementally; or when the problems you're facing are primarily executional, such as scaling infrastructure, hiring the right people, or refining processes.

These are solvable problems. They don't require abandoning the strategy. They require better execution of the existing strategy.

## THE PIVOT SIGNALS

Pivot when customers are indifferent. Pivot when growth requires constant force—when you're pushing a boulder uphill and the moment you stop pushing, it rolls backward. Pivot when your team spends more time defending the strategy than improving the product—when conversations shift from "How do we make this better?" to "Why are we still doing this?"

Pivot when you find yourself explaining away the data rather than learning from it. The discipline lies in distinguishing emotional commitment from empirical traction.

## EVIDENCE OVER EGO

Great leaders don't quit when things get uncomfortable. Discomfort is normal. Building anything worthwhile is hard. But great leaders do quit when reality makes their original plan indefensible. This is adaptive intelligence at its core: the ability to hold conviction lightly while holding evidence tightly.

You believe in your vision—but not more than you believe in observable truth. You're committed to your mission—but flexible on the path to get there. You have the confidence to start and the humility to change direction when the market tells you something isn't working.

Ben Horowitz faced this at Loudcloud. The hosting business wasn't just struggling—it was broken. Customers were failing. The model was unsustainable. He could have told himself that persistence meant fighting harder for the original vision.

Instead, he looked at the evidence. The software was working. The hosting wasn't. The choice was clear.

Kevin Systrom and Mike Krieger faced this with Burbn. They could have kept adding features, kept trying to make check-ins work, and kept telling themselves that persistence meant sticking with the original plan. Instead, they looked at the data. Photos were working. Check-ins weren't. The choice was clear.

George Moye III faced this at Guilford Courthouse. He could have interpreted Greene's retreat order as cowardice and as a failure of will. Instead, he understood the strategic reality. Holding ground wasn't winning. Preserving the army to fight another day was winning. The choice was clear.

The pattern is consistent: evidence over ego. Mission over method. Strategic clarity over tactical stubbornness. That's the difference between necessary persistence and counterproductive stubbornness. And it's the difference between leaders who adapt intelligently and leaders who drive their organizations into the ground defending indefensible positions.

The contemporary business environment makes this distinction even more critical. With the rise of AI, the speed of market disruption has accelerated dramatically. Companies that were dominant five years ago are struggling now because they fell in love with solutions that worked in 2019 but don't work in 2026.

## THE FEATURE CREEP TRAP

There's a specific failure mode that kills startups and corporate innovation projects alike: feature creep.

Feature creep happens when you keep adding capabilities to your product because customers request them, because competitors have them, or because they seem like good ideas,without stepping back to ask whether they serve your core purpose.

This is the opposite of adaptive intelligence. It's reactive addition rather than strategic focus.

Imagine if Nathanael Greene had responded to Guilford Courthouse this way:

"The militia broke faster than we wanted in the first line. We need to add more training. And we need more artillery support. And we should build earthwork fortifications. And we need cavalry to protect our flanks. And we should coordinate with local irregulars. And we need better supply lines for ammunition. And..."

He could have spent months trying to add capabilities to address every tactical shortcoming, trying to fight the British head-on with a perfectly optimized army. Instead, he accepted constraints, focused on what he could actually control, and designed a strategy that used his limitations as advantages—militia that would retreat on command, smaller numbers that could move faster, logistics that favored maneuver over pitched battles.

**Adaptive intelligence means saying no to feature additions that don't serve the core problem you're solving.**

In startups, this shows up constantly:

- Enterprise sales team wants custom features for big clients (feature creep)
- Product team wants to add capabilities competitors have (feature creep)
- Customer success team wants to build solutions for specific user complaints (feature creep)

Each individual request seems reasonable. Taken together, they create bloated products that don't do anything particularly well because they're trying to do everything adequately.

The Instagram story is instructive here too. After they pivoted from Burbn to Instagram, they had to resist constant pressure to add features:

- "Why can't we post videos?" (They eventually added this, but only after photos were definitively working)
- "Why can't we edit photos after posting?"
- "Why can't we add locations?"
- "Why can't we message people privately?"

They said no to almost everything for a long time, staying focused on making photo sharing with filters as simple and engaging as possible. Only after they'd achieved product-market fit did they carefully add features.

## THE ZOOM DISCIPLINE

Consider Zoom. It's impossible to think about modern work without Zoom—a company name now synonymous with video conferencing, the way Xerox became synonymous with photocopying in the 1980s. The rise of Zoom wasn't an accident of timing. It was the result of unusually disciplined product leadership by founder and CEO Eric Yuan.

When Yuan started Zoom, the video-conferencing market was already crowded. WebEx, Skype, and others had been around for years. They were feature-rich, offering document sharing, screen control, whiteboarding, chat integration, workflow tools—everything an enterprise could want. And they were frustrating to use. Calls dropped. Video froze. Audio cut out. Meetings started late because half the participants couldn't get the software to work.

Yuan made a counterintuitive decision: Zoom would obsess over one thing—delivering consistently high-quality video that worked reliably under real-world conditions. That meant investing deeply in network optimization, compression algorithms, latency reduction, and graceful degradation on poor connections—the invisible engineering details that most users never notice when they work, but immediately curse when they fail. Internally,

Zoom engineers weren't measured on how many features they shipped. They were measured on whether meetings started on time, whether audio didn't drop, and whether video didn't freeze. Reliability, not breadth, became the company's defining metric.

Here's the crucial part: Yuan said no to a long list of features that customers and enterprises routinely requested. Early Zoom intentionally avoided becoming a full collaboration suite. It resisted deep document sharing. It deprioritized complex workflow tools that competitors marketed aggressively. The logic was simple but rare: Every additional feature risked introducing friction, instability, or cognitive overhead that would undermine the core promise—that the call just works.

Yuan understood something fundamental about video software: Users don't praise it when it works. They complain only when it fails. By focusing relentlessly on the invisible engineering that makes quality feel effortless, Zoom earned trust at scale. That trust, once established, became the platform on which later features could safely be added—but only after product-market fit was undeniable.

This echoes one of Steve Jobs's most famous leadership principles at Apple: "Focus is about saying no." Jobs clarified that focus doesn't mean saying no to trivial distractions. It means saying no to a thousand good ideas to protect the great one. Zoom embodies that principle in modern software form.

Yuan's leadership demonstrates that focus isn't a lack of ambition—it's a sequencing strategy. Earn the right to expand by first becoming indispensable at something narrow and critical. Don't try to do everything. Do one thing so well that people can't imagine working without it. Then, and only then, carefully add capabilities that enhance rather than dilute your core value.

This is the same discipline Nathanael Greene showed at Guilford Courthouse. He didn't try to match the British in every dimension. He didn't keep adding capabilities until he had a conventional army that could fight conventionally. He accepted what he couldn't control. He focused on what he could. He designed a strategy that turned limitations into advantages. The breakthrough came not from building more, but from choosing what not to build. That's adaptive intelligence.

## THE MINIMUM VIABLE PRODUCT DOCTRINE

This connects to one of Silicon Valley's most important frameworks: the Minimum Viable Product (MVP).

The MVP concept—popularized by Eric Ries in *The Lean Startup*—says: Build the simplest version of your product that lets you test your core hypothesis, then iterate based on what you learn. But most people get MVPs wrong. They think "minimum" means "cheap and quick." It doesn't. It means "minimum *viable*"—the smallest thing that actually solves the core problem for early users well enough that they'll use it and give you feedback.

Instagram's MVP was photo sharing with filters. That's it. No videos. No direct messages. No stories. No reels. Just photos with filters. But those photos with filters worked *really well.* The filters were beautiful. The sharing was instant. The experience was delightful. It solved the core problem—helping people share visually appealing moments—completely, even though it didn't do a hundred other things people eventually wanted.

**This is the Guilford Courthouse strategy applied to product development: Design the minimum engagement that achieves your strategic objective, execute it with discipline, learn from the results, and then iterate.**

Greene didn't try to win the entire war at Guilford Courthouse. He tried to weaken Cornwallis enough to make the British campaign unsustainable. That was his MVP—and it worked.

But here's where leaders struggle: How do you distinguish a true Minimum Viable Product from a merely incomplete product? The difference is subtle but decisive. A true MVP feels narrow but complete. An incomplete product feels unfinished and apologetic.

The most useful starting point is defining the product's single irreplaceable job. Leaders should ask: If this product disappeared tomorrow, what specific pain would users immediately feel again? If the answer is vague—"productivity," "collaboration," "engagement"—the MVP is not yet defined. When Zoom launched, its job was crystal clear: Enable a video call that starts instantly and doesn't fail. That clarity allowed Eric Yuan to reject features that did not directly strengthen that job. A product is viable when it solves its primary job end-to-end without workarounds, training, or user forgiveness.

The second lens is trust completeness, not functional completeness. Product leaders should ask: Does this product make users confident enough to rely on it again tomorrow? Viability is achieved when users trust the product to perform consistently in its core scenario. An MVP can lack edge cases, customization, and secondary workflows, but it cannot fail at the moment of truth.

This is why many early products fail despite technically shipping. They collapse under real-world usage. A half-working core is not an MVP; it's reputational debt. Users don't forgive products that fail at the one thing they're supposed to do well. This principle echoes Steve Jobs's insistence that products feel whole, even when limited. Apple's early products often had fewer features than competitors, but what they did have worked flawlessly, reinforcing trust rather than testing patience.

## SAYING "NOT YET" VS. "NEVER"

A third question separates discipline from desperation: Are we saying "not yet" to features, or "never"? True MVP thinking includes an explicit roadmap of deferred capabilities, with a rationale tied to learning sequence. Leaders should be able to articulate why each excluded feature is postponed and what evidence would justify its inclusion later.

This prevents false minimalism, where teams underbuild simply to ship faster without strategic intent. Jobs captured this distinction when he said that focus means saying no to a thousand things—not because they're bad ideas, but because they're out of sequence. A product is incomplete when it lacks intent. It is minimal when it reflects deliberate prioritization.

Finally, leaders should test MVP viability with behavioral proof, not verbal feedback. Ask: Are users returning unprompted? Are they adapting their workflows around the product? Are they recommending it without incentives? These signals matter more than feature requests, which often reflect users projecting existing habits onto a new tool rather than responding to what the product actually delivers.

A true MVP creates pull. It changes behavior with minimal explanation. An incomplete product requires persuasion, excuses, and constant hand-holding. The difference between the two is leadership judgment: the courage to delay breadth, protect coherence, and ship only what makes the product unmistakably valuable on day one.

This is exactly what Greene did at Guilford Courthouse. He didn't apologize for having militia instead of regulars. He didn't try to convince anyone that his smaller force was "almost as good" as the British army. He designed a battle plan that made his constraints irrelevant by focusing relentlessly on the one thing he could do better than Cornwallis: Move fast, strike hard, and preserve his force to fight again. The plan was minimal. But it was complete. And it worked.

## ITERATION VS. PIVOT: KNOWING THE DIFFERENCE

There's an important distinction here between iteration and pivoting:

- **Iteration** means you're making the same basic solution better—refining, improving, optimizing.
- **Pivoting** means you're changing the fundamental solution while staying focused on the same problem.

Instagram didn't iterate Burbn into Instagram. They pivoted—they fundamentally changed what they were building while staying focused on the same problem (helping people share moments).

Nathanael Greene didn't iterate conventional military tactics into slightly better conventional tactics. He pivoted—he fundamentally changed how he fought while he stayed focused on the same problem (defeating the British in the South).

Many leaders fail because they iterate when they should pivot, or pivot when they should iterate.

**Iterate when:**

- The core solution is working but needs refinement
- User feedback is about improving features, not questioning the whole approach
- Metrics are trending positive but not as fast as you'd like
- The problem is execution, not strategy

**Pivot when:**

- The core solution isn't creating the engagement or value you expected
- User feedback suggests you're solving the wrong problem

- Metrics are flat or declining despite execution improvements
- The market has shifted in ways that make your current approach obsolete

## ITERATE VS. PIVOT—FOUNDER DECISION TABLE (MINIMAL AND ACTIONABLE)

| **Signal** | **ITERATE (Stay the Course)** | **PIVOT (Change Direction)** |
|---|---|---|
| Retention | Cohort retention is flat or improving (≥25–30% consumer, ≥40% B2B) | Retention collapses after first use (<10–15%) |
| Core Usage | Same core feature used repeatedly | Usage is shallow or scattered |
| User Pull | Organic referrals or unprompted reuse | Growth requires constant push or incentives |
| Willingness to Pay | Some users pay or try hard to justify paying | No one pays without deep discounts |
| Effect of Iteration | Shipping fixes moves engagement metrics | Shipping fixes changes nothing |

Greene faced this decision after Camden. He could have iterated—taken Gates's approach and just executed it better. Instead, he recognized that conventional tactics wouldn't work against Cornwallis with the forces he had, so he pivoted to a strategy of maneuver and calculated engagements.

## THE AI DISRUPTION MOMENT

We're living through a pivot moment right now with artificial intelligence. GPT-3.5 launched in November 2022, and ChatGPT made AI capabilities accessible to ordinary users for the first time. Within months, the technology landscape shifted more dramatically than it had in decades.

Companies are facing the same choice George Moye III faced at Guilford Courthouse: Do we stand and fight with our current approach, or do we execute a calculated retreat and redeploy?

Many established companies are trying to iterate—adding AI features to existing products, rebranding existing capabilities as "AI-powered," and basically trying to compete with new AI-native companies using legacy architectures.

**This is like trying to win Guilford Courthouse using Camden's tactics. It won't work.**

The companies that will win are those willing to execute Nathanael Greene's strategy: Acknowledge that the old approach is obsolete, preserve your core assets (brand, customer relationships, and distribution), and redeploy to fight where the new technology gives you an advantage.

When ChatGPT launched in late November 2022, Google had everything it needed to respond decisively. Superior language models. Deeper AI research than almost any company on earth. Armies of talented engineers who'd literally invented the transformer architecture that made ChatGPT possible. They should have been able to move fast.

They didn't.

For months, Google hesitated. Not because they lacked the technology. Not because they didn't understand the threat. They hesitated because ChatGPT threatened to make their entire business model obsolete.

Google had built a $160 billion advertising empire on a specific paradigm: You search for something, Google shows you ten blue links with ads mixed in, you click, and Google gets paid. The whole infrastructure—the algorithms, the metrics, the quarterly targets, and the sales compensation—reinforced that model. It had made them one of the most valuable companies in history. And now a conversational AI was offering to answer questions directly, no links required. No clicks meant no ads. No ads meant no revenue model.

Google wasn't just competing with a new product. They were fighting their own muscle memory. Every system they'd built over twenty years pushed them toward optimizing search results, not replacing the search entirely.

When they finally launched Bard in February 2023, it was rushed and plagued with errors. In the demo, Bard gave a factually incorrect answer about the James Webb Space Telescope—and Google's stock dropped $100 billion in a single day. The infrastructure that had made them dominant had calcified into constraint.

Microsoft, meanwhile, moved faster. Not because they were smarter or had better AI researchers. They moved faster because they had less to lose. Their search business was a rounding error compared to Google's. They could integrate OpenAI's technology into Bing without cannibalizing a cash-printing empire. (Google has since rebounded considerably with their nearly infinite resources and closed much of the early AI gap—but those first months revealed something important about organizational adaptability.)

Enterprise software giants faced a parallel trap, though their response took a different form: AI-washing.

Within months of ChatGPT's launch, every major enterprise software company announced AI features. Salesforce unveiled "Einstein GPT." SAP touted "AI-powered insights." Dozens of legacy platforms rushed to slap conversational interfaces onto their existing products. But bolting a chatbot onto a CRM system designed in 2005 doesn't constitute transformation. It's a fresh coat of paint on rotting infrastructure.

These companies understood that Wall Street and customers demanded an AI story. But rebuilding from scratch meant cannibalizing existing revenue, retraining sales teams, and admitting their core products were aging. The adaptive intelligence required wasn't technological; most of them had the resources to hire brilliant AI engineers. It was strategic: the willingness to declare their own products obsolete before the market did.

Meanwhile, Notion and Linear and similar companies—unburdened by legacy codebases and shareholder expectations tuned to old business models—built AI-native products from scratch. They didn't add AI features to existing products. They reimagined what the products should be if they designed them today, with AI as a foundational assumption rather than a bolt-on enhancement.

And they captured the next generation of users while incumbents debated migration strategies in conference rooms.

## THE GUILFORD COURTHOUSE QUESTION

This is the organizational version of the question George Moye III faced at Guilford Courthouse: Do you defend the ground you're standing on, or do you execute a strategic retreat to fight on better terms?

Google's initial paralysis wasn't stupidity. It was the rational response of an organization optimized for a specific battlefield, trying to decide whether to abandon that battlefield voluntarily. The enterprise software companies' AI-washing wasn't dishonesty—it was the instinct to defend existing revenue streams rather than replace them.

But adaptive intelligence requires seeing past the immediate tactical loss to the strategic necessity. Sometimes you have to abandon the position you've been defending—even if you're currently winning there—because the terms of engagement have fundamentally changed.

Greene understood this at Guilford Courthouse. He could have tried to hold ground against the British, defending the field in conventional terms. Instead, he pre-declared a retreat, fought just enough to inflict damage, and preserved his force for future engagements on more favorable terms.

The question for every established company facing disruption is whether they can execute that kind of strategic retreat while they still have the resources and positioning to do it effectively—or whether they'll wait until the market forces the retreat on unfavorable terms.

The pattern is clear: Companies that stay in love with their legacy solutions (their "Burbn") will lose to companies that stay in love with the problem and build new solutions optimized for the AI era (their "Instagram").

## DATA: THE NEW AMMUNITION

There's another parallel worth exploring between Guilford Courthouse and modern product development. Greene's strategy depended on intelligence—knowing where Cornwallis was, how strong his forces were, how supplied his troops were, and what routes he was likely to take. The more information Greene had, the better he could design engagements that favored his forces.

Modern product development is the same, except the "intelligence" is usage data.

The Burbn founders didn't just guess that photo filters were working better than check-ins. They looked at their analytics and saw that photo engagement was dramatically higher. They had data that told them where users were finding value. This is why modern adaptive intelligence requires instrumentation—you need to see what's actually working, not just what you think should work.

Too many leaders make decisions based on opinions, intuition, or what competitors are doing, rather than look at their own data about what their users actually value. However, it is critical to be able to understand the difference between signal and noise in your data.

## THE VANITY METRICS TRAP

The seduction of dashboards is real. Founders, desperate for evidence that their startup is working, build elaborate analytics systems that track everything: page views, signups, downloads, session duration, and feature adoption rates. The data pours in. The charts trend upward. The team celebrates. And then the business fails despite "strong metrics."

The problem isn't a lack of data. It's measuring vanity instead of truth.

Netflix learned this the hard way in their early streaming days. They initially celebrated when users clicked play on a movie. The metric looked great—millions of plays per day. Engagement was through the roof. But that metric was meaningless. What mattered was whether people watched past the first ten minutes. A "play" might mean someone immediately hated their recommendation and bounced. The click told them nothing about satisfaction or content quality. Completion rate told the real story—whether users actually valued what they were watching enough to finish it.

That shift in focus from "engagement" to "valuable engagement" changed how Netflix evaluated content investment. It reshaped their recommendation algorithm. It ultimately defined how the platform works today. The lesson is universal: The metrics that make you feel good rarely align with the metrics that predict survival.

## LEADING VS. LAGGING INDICATORS

Distinguishing signal from noise requires understanding the difference between leading indicators and lagging indicators. Most founders confuse the two.

Take the daily active user (DAU)—the metric every consumer startup obsesses over. DAU is a lagging indicator. By the time it drops, your product has already been dying for weeks. Users have already decided they don't need you. You're just now seeing the numbers reflect that reality.

The leading indicator is more subtle: day-7 retention among users who completed onboarding in the past week. If that number softens, you're watching users decide in real time that your product isn't essential to their lives.

Duolingo discovered this when they noticed their day-1 retention was strong—over 80 percent. But day-7 retention collapsed to 15 percent. Something was happening in that first week that made users quit. The aggregate DAU trend wasn't helpful. The signal was buried deeper—in the behavioral cohorts of users who completed exactly three lessons on day one versus those who completed one or five.

Three lessons hit what they called a "competence moment." Users felt genuine progress and believed they could actually learn the language. One lesson was too shallow to matter. Five lessons created intimidation; it felt like homework.

By optimizing around that insight—gamifying the path to three lessons and making it feel natural and achievable—Duolingo transformed retention across their entire user base. The noise was the overall DAU trend going up or down. The signal was in granular behavioral patterns that revealed what actually drove users to come back.

## THE SURVIVORSHIP BIAS TRAP

The most dangerous mistakes in interpreting usage data stem from ignoring who isn't in your dataset: the survivorship bias trap.

When Superhuman analyzed their feature usage data, they initially celebrated. Eighty-five percent of active users loved their blazing-fast keyboard shortcuts and advanced email triage features. The product worked. Users were delighted. Product-market fit achieved.

Then they ran a haunting experiment: They surveyed users who had signed up but never activated their accounts. The feedback was unanimous: "Too complicated," "Felt like homework," "Overwhelming." The 85 percent satisfaction score wasn't measuring product-market fit. It was measuring power-user fit. They'd built a product that delighted a narrow segment of email power users while alienating everyone else. But their data captured only the survivors—the people who'd stuck around long enough to become power users themselves.

This pattern shows up everywhere. Many SaaS companies make pricing decisions based on average revenue per user (ARPU), only to discover that "average" masks a critical truth: 80 percent of revenue comes from 20 percent of customers with wildly different needs. Spotify's average user might stream 15 hours per month, but that average combines casual listeners streaming 3 hours with superfans streaming 60+ hours. Those segments have completely different values and completely different churn triggers. Optimizing for the average user means building a product that perfectly serves no one.

## THE FEW METRICS THAT MATTER

The path to better data-driven decisions isn't more metrics. It's relentless focus on the few that actually predict outcomes.

Jeff Bezos famously decreed that Amazon would obsess over three numbers: customer acquisition cost, lifetime value, and free cash flow per share. Everything else was context, not core. For early-stage startups, the equivalent holy trinity is even simpler.

**First: Retention curve shape.** Does it flatten or race to zero? A flattening curve means you've found something people genuinely need. A curve that races to zero means you haven't, no matter how good your other metrics look.

**Second: Time-to-value.** How long until a new user experiences the core benefit? Dropbox knew they'd won when they reduced this from "days until your second device syncs" to "seconds until your first file uploads." The faster users experience value, the more likely they are to stick around.

**Third: Frequency of naturally occurring use.** Are users returning because your push notifications nag them into opening the app or because they have an organic need that your product fulfills? Slack's breakthrough metric was "2,000 team messages sent." Teams that hit that threshold almost never churned, because the communication habit had formed. The product had become essential to how they worked.

Notice what's absent from this list: total users, press mentions, app store rankings, and social media followers. Those metrics feel good. They make for great board meeting slides. They impress investors who don't look closely. But they predict nothing about whether your business will survive. The hardest discipline in data-driven product development isn't building better dashboards. It's having the courage to ignore 95 percent of what those dashboards tell you and bet everything on the 5 percent that actually matters.

This is adaptive intelligence applied to metrics: Know what you're really trying to measure, distinguish signal from noise, and have the discipline to focus relentlessly on the indicators that predict survival rather than the ones that make you feel successful.

Greene had scouts, informants, and captured intelligence that told him Cornwallis's situation. The modern equivalent is analytics dashboards, user feedback, A/B testing, and cohort analysis. The companies that build this intelligence-gathering capability—that know exactly what features drive engagement, what user journeys lead to conversion, and what behaviors predict retention—can adapt faster than companies operating blind.

## PRESERVING YOUR FORCE

There's one more critical lesson from Guilford Courthouse that applies directly to modern leadership.

Greene's primary objective wasn't killing British soldiers. It was preserving the Continental Army as a fighting force. If he'd stood and fought to the last man at Guilford Courthouse, he might have inflicted more British casualties. But he would have lost his army—and with it, any chance of continuing the campaign. By retreating when necessary, he kept his army intact. They could fight again. They could maneuver. They could eventually win the war.

**This is crucial for startups and corporate innovation projects: Your most important asset is your team and everyone's morale.**

When a product isn't working or when a strategy is failing, the wrong response is to keep pushing until everyone burns out and quits. The right response is to preserve your team's energy and enthusiasm by pivoting before you've exhausted everyone.

I've seen this pattern repeatedly:

- Startup A keeps fighting for a failing product until the team is burned out, relationships are fractured, and everyone wants out. When they finally pivot, they've lost their best people.
- Startup B recognizes earlier that their approach isn't working. They pivot while the team still has energy. They reframe it as learning rather than failure. The team stays together and excited.

Guess which startup has a better chance of succeeding on their second try?

Greene preserved his force. The Instagram founders preserved their team's morale by pivoting before everyone gave up. Smart leaders preserve their most valuable asset—committed, capable people—by adapting before that asset is depleted.

## THE TRANSPARENCY PRINCIPLE

The worst thing a founder can do during a struggling pivot isn't admitting the product is failing. It's pretending it isn't.

When Stewart Butterfield gathered his Tiny Speck team to announce they were shutting down Glitch after years of work, he didn't sugarcoat the news or hide behind spin. He told the truth: the game wasn't working. They'd run the experiments. They'd given it every reasonable chance. Continuing would mean slowly dying while burning cash and morale. Then he did something unexpected. He asked the team what they'd noticed. The developers admitted they'd spent more time perfecting their internal chat tool than game mechanics. The designers confessed they were more excited showing the communication interface to friends than the game itself.

By making the team co-authors of the pivot narrative rather than victims of an executive decision, Butterfield transformed potential resentment into ownership. The transparency created trust: If leadership was honest about failure, they could be trusted about the next direction. The team didn't feel betrayed. They felt respected.

Contrast this with Quibi, where leadership spent months insisting their short-form video platform just needed "one more quarter" while employees watched metrics crater and began quiet quitting months before the official shutdown. The team knew it was over. Leadership's refusal to acknowledge reality didn't preserve morale—it destroyed credibility. Employees lost faith not because the product failed, but because leadership couldn't admit what everyone already knew.

The teams that survive pivots share a common trait: They move with conviction rather than drifting through uncertainty.

Instagram's pivot from Burbn—a cluttered check-in app with photo features—to a pure photo-sharing app took one weekend. Kevin Systrom and Mike Krieger didn't spend months in analysis paralysis. They didn't try to A/B test their way to clarity. They looked at their data—photos had 10 times the engagement of check-ins—made the call, and rebuilt. The speed mattered as much as the decision itself. Prolonged ambiguity is team morale poison. When everyone knows something isn't working but leadership won't decide, the best employees start interviewing elsewhere while the company hemorrhages away.

Twitter's pivot from Odeo, a podcasting platform about to be obliterated by Apple's iTunes, to microblogging happened during a single brainstorming session when Jack Dorsey proposed status updates via SMS. The team rallied because clarity replaced confusion. They'd been watching their podcasting business die slowly. When leadership finally made a definitive call on a new direction, the relief was palpable.

Compare this to countless startups that spend six to twelve months in "pivot purgatory"—testing multiple directions simultaneously, launching half-hearted experiments, and hoping the market will tell them what to build. The hedging signals lack of conviction, and teams can smell the fear. Employees need founders who can say, "We're betting everything in this direction," not "Let's try three things and see what sticks."

Preserving team morale during pivots requires protecting two things: people's sense of competence and their belief in mission continuity.

When YouTube pivoted from a video dating site to general video sharing, they didn't tell engineers, "Your dating algorithm work was wasted." They reframed it: "You built infrastructure that can handle any video use case, which is far more valuable than we originally envisioned." The code wasn't thrown away. It was repositioned as the foundation for something bigger. The engineers who'd spent months building video processing and recommendation systems didn't feel that their work had been pointless. They realized they'd built something even more important than planned.

PayPal executed this brilliantly during their pivot from PalmPilot payments to email-based transfers. Peter Thiel didn't tell the team, "We failed at mobile payments." He celebrated that they'd built the world's most secure transaction system and cryptographic infrastructure, which would now power something with 100 times the addressable market. The pivot felt like expansion, not retreat. The team's core competency—building secure payment systems—remained valuable. They were just applying it to a bigger problem.

Contrast this with the dozens of crypto startups that pivoted to "blockchain for enterprise" during the 2018 crash. They asked true-believer engineers who'd joined for decentralization ideals to build corporate databases. The mission whiplash was complete, and no amount of free lunches could compensate for the feeling they'd abandoned their principles for survival. Many of those teams suffered spectacular talent attrition, even when the pivots worked financially. The technical work might have been similar, but the mission had evaporated.

The difference between teams that burn out forcing a failing strategy versus those that energize through a pivot comes down to psychological safety and sunk cost discipline.

Airbnb's near-death experience in 2009 offers a master class. They'd spent months building elaborate features while barely anyone booked rooms. The founders could have kept iterating on the product, adding more features, and improving the algorithm, hoping something would click. Instead, they did something radical: They went to New York, knocked on host doors, and took professional photos of listings themselves. It wasn't scalable. It wasn't "tech." It definitely wasn't what investors wanted to hear. But it worked—bookings tripled.

The team didn't burn out because leadership was willing to do unglamorous, ego-bruising work to learn the truth. They weren't attached to their vision of what Airbnb "should be." They were attached to solving the problem of making people trust strangers enough to stay in their homes. When the data showed that better photos solved that problem, they did whatever it took to get better photos, even if that meant Brian Chesky personally shooting photos with a borrowed camera.

Conversely, Theranos is the canonical example of burning out while forcing a failing strategy. Employees who raised concerns about the technology were marginalized or fired. The mission became defending the narrative rather than fixing the product. People who told the truth about what the machines could and couldn't do were treated as disloyal rather than valuable. The culture punished honesty, so honesty disappeared, and with it, any chance of adapting to reality.

Teams that survive pivots embrace what Andy Grove called "intellectual honesty"—the willingness to see the world as it is, not as you wish it were. The morale killer isn't pivoting; it's leaders who can't admit when they're wrong, and teams that are punished for telling the truth. Build a culture where bad news travels fast, and you'll have a team resilient enough to pivot. Build a culture where reality is negotiable, and you'll have a team that abandons ship the moment things get hard.

This is the organizational equivalent of what George Moye III and his fellow soldiers experienced at Guilford Courthouse. Greene's pre-declared retreat worked only because the soldiers trusted their leadership enough to execute it. If Greene had spent months denying that the Continental Army was weaker than the British, if he'd punished officers who admitted they couldn't hold ground against professional soldiers, or if he'd insisted they could win a conventional battle through sheer willpower, his army would have disintegrated.

Instead, he told them the truth: "We're outgunned, but we're not outmaneuvered. We're going to fight on our terms, not theirs. We're going to hit them hard and pull back before they can destroy us." And because his soldiers trusted that honesty, they executed the strategy with discipline.

The same principle applies to startup teams facing pivots. Tell them the truth. Move with conviction. Protect their sense that their work matters. And build a culture where reality is welcomed, not feared. That's how teams survive strategic retreats without falling apart.

## THE DISCIPLINE OF RETREAT

The hardest part of Guilford Courthouse wasn't the fighting—it was the retreating.

It's relatively easy to train soldiers to stand and fire. It's incredibly difficult to train them to retreat on command without degenerating into a panicked rout. That third volley was the key moment. After firing twice and reloading under pressure, with British bayonets closing fast, every instinct screamed: *Run now! Don't wait! Just go!*

But if they'd run then—before delivering that third, devastating volley—they would have wasted the whole strategy. The British would have taken lighter casualties. The retreat would have been disorganized. Men would have been captured or killed unnecessarily. The discipline to execute the plan when fear was overwhelming made the difference between strategic retreat and catastrophic failure.

**In business, this translates to having the discipline to execute your pivot systematically rather than panic-pivoting.**

When founders realize their product isn't working, the temptation is to immediately change everything—new features, new target market, new positioning, and new strategy—all at once. This is panic. This is the militia breaking before the third volley.

Better to:

1. Identify specifically what's not working (data)
2. Understand why it's not working (user research, market analysis)
3. Hypothesize what might work instead (product strategy)
4. Design a minimum viable test of that hypothesis (MVP)
5. Execute deliberately, preserving team morale and resources

This is disciplined retreat. This is the third volley before falling back.

A systematic pivot follows a counterintuitive sequence: Decide fast, communicate immediately, but execute in phases. The decision itself should take days, not months. Once you've identified the signal—that accidental use case, that cohort with extraordinary retention, or that adjacent market pull—prolonged deliberation only burns runway and team confidence. Stripe's founders famously made their pivot from backend payments API to developer-focused infrastructure in a 72-hour period after noticing their side project had more GitHub stars than their main product. The insight was clear. The decision was fast. The execution, however, took quarters.

The first week is about **transparency with your team**. Tell them the full truth in a single all-hands meeting. Cover why the current approach isn't working, and bring data to support it. Explain what you're pivoting to, and communicate it with conviction, not ambivalence. Most importantly, clarify what stays the same such as the mission, the values, and the fundamental problems you're trying to solve. This isn't about abandoning your purpose. It's about finding a better path to achieve it.

The second week is **investor communication**. Write a structured memo covering three elements: what you learned that necessitates the change, evidence that the new direction has traction, and an updated financial model showing the path to profitability or the next milestone. Frame this as strategic evolution based on customer insight, not panic. Investors fund teams that learn fast, not teams that never make mistakes. The founders who get crucified aren't the ones who pivot. They're the ones who hide pivots until they're out of money.

The third week is **customer segmentation**. Power users who love you get early access and co-creation opportunities. Invite them to help build the next version. Casual users get a clear transition timeline with plenty of notice. Non-engaged users get a simple sunset announcement with data export options. Notion executed this masterfully when pivoting from a consumer note-taking tool to an "all-in-one workspace." They invited their most engaged users to beta test the new features, transforming them from potential skeptics into evangelists who championed the change.

The execution timeline must balance speed with team preservation. Move fast enough that you don't die, but slow enough that you don't shatter. The mistake most founders make is trying to pivot and maintain the old product simultaneously. This splits engineering resources and guarantees mediocrity in both directions. You can't win two wars at once with limited troops. The discipline required is establishing a hard cutoff date—typically 30 to 60 days out—when all work on the legacy product stops, and communicating this internally with brutal clarity.

During those 30 to 60 days, assign a "sunset team," typically 10 to 20 percent of engineering, to maintain the old product while everyone else builds the new direction. No hedging. No "let's do both and see which one works." Make the call and commit the resources.

Slack gave themselves exactly eight weeks from pivot decision to Slack 1.0 launch, with the old game entering maintenance mode immediately. The compressed timeline forced prioritization. They shipped with a fraction of the features they'd originally envisioned, but the core value proposition—searchable, organized team communication—was crystal clear. They didn't try to rebuild everything. They built the minimum necessary to prove the concept worked and invited early users to give feedback.

For customers, provide a generous transition window but clear milestones. Set a date when the old product enters maintenance mode—no new features, just security updates and critical bug fixes. Set a sunset date, typically six months later, when the product shuts down completely. Provide clear data export and migration paths. Respect the people who trusted you enough to use your product, even if you're moving on from it.

The teams that fail at pivots typically fail at this juncture. They keep the old product on life support for 18 months while starving the new direction of resources. They're trying to avoid upsetting existing customers, but the result is that they satisfy no one. The old product stagnates and frustrates users who want improvements. The new product can't get the resources it needs to succeed. The company burns through the runway without making meaningful progress in either direction.

The hardest moment in executing a pivot isn't making the decision. It's not communicating it to your team, your investors, or your customers. The hardest moment is having the discipline to truly let go of what you built. You spent months or years building the original product. You believed in it. You convinced other people to believe in it. You raised money for it. You hired people to build it. And now you're killing it. That requires grief. Let yourself feel it. Give it a day. Acknowledge what you learned, what you built, and what didn't work. Then sprint toward the new direction with the same intensity you had on day one of the company.

This is the organizational equivalent of what Nathanael Greene did at Guilford Courthouse. He didn't keep half his army defending the field while sending the other half to retreat. He committed to the retreat. He executed it with discipline. He preserved his force to fight the next battle on better terms. The pivots that succeed are the ones where leadership makes a clear decision, communicates it honestly, and commits resources decisively. The pivots that fail are the ones where leaders hedge, split resources, and can't bring themselves to fully abandon what they've built.

Adaptive intelligence isn't just about knowing when to pivot. It's also about having the courage to execute the pivot completely once you've made the decision.

## WHEN PURPOSE ENABLES ADAPTATION

There's a connection here to the themes from Section I about purpose.

George Moye III could execute Greene's complex retreat strategy because he understood the larger purpose: not winning this battle, but winning the war. That purpose freed him to adapt his tactics without feeling that he was abandoning his mission. If his purpose had been "hold this specific piece of ground" or "never retreat," he couldn't have adapted. The wrong purpose constrains adaptation.

This happens in companies all the time:

- Company purpose: "Be the leading provider of on-premises enterprise software" → Can't adapt to cloud computing
- Company purpose: "Maximize quarterly earnings" → Can't make long-term investments required to pivot
- Company purpose: "Defend our market share in X" → Can't cannibalize X to move to Y, even when Y is clearly the future

But if your purpose is framed around the problem you solve rather than your current solution:

- "Help businesses manage customer relationships effectively" → Can pivot from on-prem to cloud to AI-native
- "Create long-term value" → Can sacrifice short-term earnings to fund necessary pivots
- "Serve our customers' evolving needs" → Can cannibalize your own products when customer needs shift

**Purpose either enables adaptation or constrains it, depending on how it's framed.**

The companies that survive disruption are those whose purpose is tied to customer problems rather than current solutions. This lets them pivot without losing identity.

Consider Microsoft under Steve Ballmer versus Microsoft under Satya Nadella. The difference illustrates how purpose can either cage or liberate a company.

Ballmer's Microsoft defined itself as "Windows and Office everywhere"—a purpose so product-specific that it blinded them to paradigm shifts. When the iPhone launched in 2007, Ballmer famously laughed it off because it didn't run Windows. When cloud computing emerged, Microsoft fought it because Azure would cannibalize Windows Server licenses. Every strategic decision had to serve Windows dominance, even as the world moved to mobile and cloud. The purpose was a straitjacket.

Nadella's first act as CEO in 2014 was rewriting the mission: "Empower every person and organization on the planet to achieve more." Notice what disappeared from that statement. Any mention of Windows. Any mention of Office. Any mention of specific products. The new purpose was platform-agnostic, which gave Microsoft permission to do things that would have been heretical under Ballmer. They built iOS and Android apps that worked better than Windows Phone apps. They are open-sourced .NET. They treated Windows as one distribution channel among many rather than the sacred center of the universe.

The Azure cloud business, now worth over $100 billion annually, became possible only when leadership admitted that Windows wasn't the future. The purpose shift wasn't feel-good corporate speak. It was demolishing the psychological barriers that prevented adaptation.

IBM offers the counter-narrative—a company that has survived multiple extinction-level technology transitions precisely because their purpose was never about the technology itself. IBM pivoted from tabulating machines to mainframes to PCs to enterprise services to cloud and AI. Each transition required cannibalizing the previous cash cow. Each time, industry analysts predicted IBM's death. Each time, IBM survived by abandoning what they'd been selling and rebuilding around what customers actually needed.

How did they do it? Their purpose, articulated as early as the 1960s, centered on solving complex business problems through information technology—not selling specific machines and not dominating specific markets, but solving business problems, using whatever technology worked best.

When Lou Gerstner took over a dying IBM in 1993, he didn't pivot to a new purpose. He returned to the original one. He shut down the PC manufacturing business that had defined IBM in the public mind for a generation and rebuilt the company around enterprise services and software. The decision looked insane to outsiders—IBM invented the PC, and now they were abandoning it? But the purpose—"solve complex business problems"—gave them permission to be product-agnostic. If PCs weren't the best way to solve enterprise problems anymore, stop selling PCs.

Contrast IBM with Kodak, whose purpose became inextricably tangled with film. Kodak invented the digital camera in 1975. Steven Sasson, a Kodak engineer, built a working prototype that captured images without film. Leadership's response was to suppress it. Digital photography threatened the film business that generated 70 percent of Kodak's profits. The film business wasn't just their revenue source. It was their identity.

Their stated mission was "help people capture and share memories." That purpose should have made the transition to digital obvious. Helping people capture memories has nothing to do with the substrate those memories are stored on. Film or digital, the purpose remains the same.

But the unstated purpose—the real purpose that drove decisions—was actually "sell film and chemicals." When executives had to choose between the stated purpose and the profit center, the profit center won every time. By the time Kodak fully committed to digital in the 2000s, smartphones had already commoditized the camera. They'd waited too long, constrained by a purpose that looked customer-focused on paper but was actually product-focused in practice. They filed for bankruptcy in 2012.

Nokia suffered an identical fate. They invented the smartphone concept with the Nokia 9000 Communicator in 1996—a device with email, web browsing, and apps, years before the iPhone. They had the technology. They had the market position. They had everything they needed to dominate the smartphone era. But they couldn't pivot away from their hardware-centric identity. Their purpose had been centered around "connecting people through mobile devices we manufacture" rather than "connecting people, period."

When Stephen Elop took over as CEO in 2010, he wrote the famous "burning platform" memo, acknowledging that Nokia was in crisis. But it was too late. If their purpose had truly been "connecting people" without the hardware constraint, they might have become a software platform company. They might have built an operating system that competed with iOS and Android. They might have pivoted to services and apps.

Instead, they sold to Microsoft in 2013 for $7.2 billion, a fraction of their peak value of over $300 billion. They were constrained by a purpose that had become their tomb. The pattern is consistent across all these examples. Companies survive paradigm shifts when their purpose is broad enough to permit adaptation. They die when their purpose becomes synonymous with their current products.

Microsoft under Ballmer couldn't adapt because "Windows and Office everywhere" locked them into specific technologies. Microsoft under Nadella could adapt because "empower every person and organization" permitted any technology that served that goal. IBM survived a century of technological change because "solve complex business problems" never specified how those problems should be solved. Kodak died because their real purpose—sell film—couldn't survive the shift to digital, even though their stated purpose could have.

This is adaptive intelligence at the organizational level. The question isn't whether your market will change—it will. The question is whether your purpose gives you permission to change with it, or whether it traps you in defending yesterday's solutions to yesterday's problems.

## THE LEADERSHIP CHALLENGE: TEACHING RETREAT

Here's what makes adaptive intelligence so difficult to execute: You have to teach people that strategic retreat is different from failure.

Our entire culture celebrates persistence, determination, and never giving up. We tell stories about people who succeeded because they refused to quit. We admire stubbornness. And sometimes that's right. Sometimes persistence pays off. But sometimes the intelligent choice is to recognize when your current approach won't work and pivot to something better. And our cultural programming makes that incredibly difficult.

Greene had to teach his militia that retreating on command was success, not cowardice. That took preparation, communication, and trust.

Modern leaders face the same challenge:

- When you tell your team, "We're pivoting away from the product we've been building," some people hear, "We failed and we're giving up."
- When you tell investors, "We're changing our strategy," some hear, "We didn't know what we were doing."
- When you tell customers, "We're discontinuing this product to focus on something new," some hear, "You're abandoning us."

The leadership task is reframing strategic retreat as intelligent adaptation:

- To your team: "We learned our initial hypothesis was wrong. That's the point of testing. Now we're going to build something better based on what we learned."
- To investors: "We preserved our runway and our team by pivoting before we burned out. That gives us multiple attempts to find product-market fit.
- To customers: "We're focusing our resources where we can serve you better. This change lets us build something you'll actually use instead of maintaining something you don't."

Greene communicated this to his officers and through them to the troops. The Instagram founders communicated it to their investors and early users. Smart leaders communicate it constantly, before and during the pivot, so nobody's surprised.

## THE THIRD VOLLEY MINDSET

Let me return to that moment at Guilford Courthouse—British bayonets at thirty yards, every instinct screaming to run, and George Moye III and his fellow militia soldiers holding fire just long enough to reload and deliver that devastating third volley.

That moment embodies adaptive intelligence:

- **Strategic flexibility:** Willing to retreat when it serves the larger purpose
- **Tactical discipline:** Executing the plan precisely, even under extreme pressure
- **Purpose clarity:** Understanding why this specific action matters

- **Trust:** Believing that the leader's strategy is sound, even when it feels wrong

When you're leading a pivot—when you're changing direction, abandoning work, and admitting the current approach isn't working—you need your team to deliver that third volley.

You need them to execute the current plan professionally, even while preparing to shift to the new plan. You need them to finish what needs finishing, document what needs documenting, and extract lessons from what didn't work. If your team panics and abandons everything immediately, you lose valuable intelligence and leave customers stranded. But if they can execute that disciplined transition, you preserve options and maintain credibility. This is the test of real leadership during pivots: Can you keep your team performing at a high level while simultaneously acknowledging that you're changing course?

## THE CLARITY IMPERATIVE

The cardinal sin of pivot management is creating ambiguity about individual roles while simultaneously demanding excellence in current work. This is the organizational equivalent of asking soldiers to fight with discipline while refusing to tell them whether they'll be in the next battle. It doesn't work.

When Twitter pivoted from Odeo to microblogging, Evan Williams did something counterintuitive—he gave every team member a binary choice within 72 hours: Either commit fully to building the new Twitter product, or take a generous severance package and leave with his blessing and a strong recommendation. No "wait and see." No "we'll figure out your role later." No six-month limbo in which people half-worked on a dying product while they anxiously scanned LinkedIn.

The clarity was brutal but compassionate. People who stayed knew they had a role in the future. People who left weren't trapped in a sinking ship out of politeness or inertia. Everyone could make an informed decision about their career with full information, rather than guessing whether they'd have a job in three months.

Contrast this with Yahoo's endless "strategic reviews" under Marissa Mayer, where teams spent years not knowing if their products would sunset next quarter. Performance collapsed—not because people were lazy or incompetent, but because the psychological cost of investing energy into potentially doomed work became unbearable. Why perfect a feature for a product that might not exist in 90 days? Why debug code that might be deleted next month? The ambiguity itself became the performance killer.

## CREATING TRANSITION INCENTIVES

The tactical fix is creating a hard date—30, 60, or 90 days out—when roles will be finalized and tying short-term incentives to that date. At Slack, Stewart Butterfield established a "launch bonus" payable to everyone who stayed through the Slack 1.0 release, regardless of whether they'd be on the team afterward.

This solved the premature disengagement problem elegantly. People had financial motivation to execute well on the transition, even if their long-term future was uncertain. The company got the focus and effort it needed during the critical pivot period. The employees got compensation that acknowledged the difficulty of working through uncertainty. Everyone's incentives aligned around making the transition successful.

## DIGNIFYING SUNSET WORK

The second tactical lever is elevating the meaning of "sunset work" so it doesn't feel like pushing a corpse uphill. When Airbnb decided to shut down their high-end luxury rental service, Airbnb Luxe, they didn't tell that team to "keep this alive while we figure out your next role." Instead, they reframed the mission: "You're responsible for dignified customer offboarding and data migration that protects our brand reputation. Screw this up, and you damage trust for the entire company."

Suddenly the work mattered, just differently. They assigned their best program manager to lead the wind-down. They gave the team a clear 90-day timeline with specific milestones. They celebrated the successful completion with the same fanfare as a product launch. Every customer got personal outreach with seamless refunds or transitions. The team took pride in professional execution rather than feeling that they were managing a failure.

This prevents the "checked-out zombie" phenomenon where people on sunset teams do the bare minimum because they feel disposable. If leadership treats sunset work as important, the team will too. If leadership treats it as an afterthought, performance will reflect that.

## PERFORMANCE MANAGEMENT DURING TRANSITIONS

The performance management framework during pivots must explicitly measure two things: execution quality on legacy work AND active contribution to the new direction, even if that contribution is indirect.

HubSpot handled this well when pivoting from basic marketing tools to a full CRM platform. Engineers maintaining the old email tools weren't sidelined or treated as legacy staff. They were invited to weekly "new product councils" where they could influence architecture decisions for the new CRM. This gave them a psychological investment in the future, even while their current code was being deprecated.

The message was clear: Your current work has a dignified end date, and your next work has already begun. You're not maintaining a dying product in isolation while the "real team" builds the future. You're part of the transition, and your input matters.

## TREATING PEOPLE LIKE PROFESSIONALS

The worst pivot management treats people like interchangeable resources to be "reallocated." Engineers are moved from Project A to Project B like furniture being rearranged. The assumption is that smart people can work on anything, so just reassign them and move on.

The best pivot management treats people like professionals who deserve clarity, respect for their current contributions, and transparent communication about their future. It acknowledges that pivots are difficult—emotionally, professionally, and psychologically. It provides clear timelines, honest assessments of where people fit in the new direction, and dignified exits for those who don't.

This is the team management equivalent of what Nathanael Greene did at Guilford Courthouse. He didn't lie to his soldiers about the difficulty of the situation. He didn't pretend the militia was as strong as the British regulars. He told them the truth: "We're going to fight, inflict damage, and retreat according to plan." He gave them a clear mission, honest communication about what success looked like, and leadership they could trust to tell them the truth even when it was hard.

Teams survive pivots when leadership does the same. They fall apart when leadership creates ambiguity, treats people as disposable, or expects excellence without providing clarity.

Adaptive intelligence in team management means being honest about uncertainty while minimizing it wherever possible, creating structure and incentives that align everyone's interests, and treating the work of transition—including sunset work—as valuable and worthy of respect.

## FROM GUILFORD TO SILICON VALLEY

March 15, 1781. George Moye III retreated from Guilford Courthouse with a bloodied arm, watching British soldiers claim a field littered with their own dead. To conventional observers, the battle was a defeat. The British held the field. The Americans had retreated. But George understood what had actually happened: They'd executed a strategy that looked like losing but was actually winning. They'd preserved their force while depleting Cornwallis's. They'd given up ground to gain strategic advantage.

Seven months later, Cornwallis would surrender at Yorktown. The war would be won.

**Sometimes the smartest move is the one that looks like retreat.**

In 2011, the Burbn founders abandoned their location-based check-in app and rebuilt as Instagram. To outside observers, it might have looked like failure—giving up on their product and admitting defeat. But they understood what was actually happening: They were executing a strategy that looked like losing but was actually winning. They were preserving their team's time and energy while deploying to where they had advantage.

Eighteen months later, they'd be selling to Facebook for a billion dollars.

**Sometimes the smartest pivot is the one that looks like giving up.**

The pattern holds across centuries, across contexts, and across technologies:

Adaptive intelligence isn't about stubborn persistence. It's about strategic flexibility—knowing when to hold and when to fold, when to stand and when to retreat, and when to iterate and when to pivot. It's about being in love with the problem, not the solution, so you're free to change solutions when evidence demands it. It's about preserving your force—your team, your resources, and your reputation—so you can fight again. And it's about having the discipline to execute that third volley before you retreat—to learn everything you can from your current approach before pivoting to the next one.

George III stood in that second line at Guilford Courthouse and demonstrated that sometimes the bravest decision is knowing when to pull back. The most intelligent strategy is sometimes the one that looks like failure.

Seven years later, he'd help rebuild his community, reconcile with former loyalists, and serve as sheriff—using the same adaptive intelligence in peace that he'd demonstrated in war. His leadership lesson echoes across 240 years: **True intelligence isn't rigid determination. It's flexible adaptation in service of unchanging purpose.**

The retreat from Guilford Courthouse won the Southern campaign. What will your calculated retreat enable you to win?

# CHAPTER 6

## *Street Fighting in Saigon*

January 30, 1968. Camp Bearcat, Bien Hoa Province, South Vietnam.

CAPTAIN JIM MOYE SAT in the operations briefing at Task Force Forsyth headquarters, reviewing the next day's patrol schedules. Tet—the Vietnamese lunar new year—was about to begin, and a ceasefire had been declared. Intelligence reports suggested some increased enemy activity, but nothing unusual for a major holiday.

The 2nd Battalion, 47th Infantry had spent the past months conducting riverine operations in the Mekong Delta, coordinating with Thai forces and running patrols through rice paddies and jungle. They'd become good at it—mechanized infantry operating from their M113 armored personnel carriers, moving quickly through difficult terrain, securing villages, and interdicting Viet Cong supply lines.

They were trained for rural counterinsurgency. Mobile operations. Working with local forces. Civic action programs including the orphanage project that had consumed so much of Captain Moye's time.

They were not trained for what was about to happen.

At approximately 3:00 AM on January 31, 1968, coordinated communist attacks erupted across South Vietnam. Seventy thousand Viet Cong and North Vietnamese Army troops hit more than a hundred cities, towns, and military installations simultaneously. Saigon itself—supposedly the most secure city in South Vietnam—was under assault.

By dawn, Captain Moye and the 2/47th Infantry were racing toward Long Binh Post, the massive U.S. logistics base and II Field Force headquarters about ten miles from Bearcat. Viet Cong sappers had penetrated the perimeter and blown up portions of the ammunition depot. Explosions lit the sky. Tracers arced through the darkness.

The 2/47th's M113s—designed for rice paddies and jungle trails—rumbled down Highway 15 toward a battle they'd never trained for. By February 1, they'd been redirected again: into Saigon itself, into the dense urban warren of Cholon district, where Viet Cong forces had seized buildings and were fighting from fortified positions. Captain Moye, who'd spent months learning Vietnamese phrases to talk with Buddhist monks about orphanages, was now fighting house to house in the streets of the capital city he was supposedly defending.

**Within 24 hours, everything they knew about their mission had become obsolete.**

The strategy that had worked the day before—rural patrols, civic action, and gradual pacification—was irrelevant. The enemy wasn't hiding in the jungle anymore. They were in the streets of Saigon, and they needed to be dislodged immediately.

This wasn't gradual adaptation. This wasn't iteration based on lessons learned. This was complete, overnight transformation forced by circumstances that gave zero time to prepare.

**This is what disruption actually looks like.**

## THE TET OFFENSIVE: BACKGROUND

To understand what happened to Captain Moye and the 2/47th Infantry, you need to understand what Tet represented strategically.

By late 1967, the Vietnam War had settled into a grinding pattern. U.S. forces controlled the cities and major installations.

The Viet Cong and North Vietnamese Army controlled rural areas and mountain regions. The two sides fought for the villages and countryside between—what military planners called the "war of pacification." American military leadership believed they were winning. Body counts were favorable. The Viet Cong seemed to be avoiding major engagements. Pacification programs were showing progress. General William Westmoreland, commander of U.S. forces in Vietnam, had told President Johnson and the American public that there was a "light at the end of the tunnel."

The communist leadership in Hanoi saw things differently. They understood that American public support for the war was fragile. If they could demonstrate that nowhere in South Vietnam was safe—that even Saigon could be attacked and that the U.S. Embassy itself could be breached—they could shatter the narrative of American progress.

So they planned the Tet Offensive: a coordinated, nationwide assault timed to coincide with the Vietnamese New Year, when a ceasefire was supposed to be in effect and when many South Vietnamese soldiers would be on holiday leave. The objectives weren't primarily military. The communists didn't expect to hold the cities they attacked. They expected to:

1. Demonstrate that the U.S. and South Vietnamese governments couldn't protect their own cities
2. Inspire a popular uprising in the South (this didn't happen, but it didn't matter)
3. Destroy American public confidence in the war effort (this worked spectacularly)

Tactically, Tet was a disaster for the communists. They suffered massive casualties—estimates range from 30,000 to 50,000 killed. They failed to hold any major objective. They were driven out of every city they attacked within days or weeks. But strategically, Tet was a triumph. Walter Cronkite, America's most trusted news anchor, declared after visiting Vietnam in February 1968: "It seems now more certain than ever that the bloody experience of Vietnam is to end in a stalemate."

President Johnson decided not to run for reelection. Peace negotiations began. The American war effort had been broken not by military defeat but by psychological shock.

**Tet demonstrated something crucial about warfare and competition: Tactical defeat can be strategic victory if you're fighting for different objectives than your opponent.**

For Captain Moye and the men of the 2/47th Infantry, these strategic considerations were abstract. Their immediate problem was simpler and more urgent: Viet Cong forces were in Saigon, and they needed to get them out.

## LONG BINH: THE FIRST SHOCK

The 2/47th Infantry's first objective was Long Binh Post—the sprawling logistics base that served as II Field Force headquarters and the main ammunition supply point for U.S. forces in III Corps.

Long Binh was supposed to be secure. It was massive—covering several square miles—and it was the nerve center for American operations around Saigon. If Long Binh fell or was seriously damaged, the entire III Corps operation would be compromised.

On the night of January 31, Viet Cong sappers penetrated Long Binh's perimeter and attacked the ammunition depot. They managed to detonate some of the stored munitions, creating spectacular explosions that could be seen for miles. When the 2/47th Infantry arrived, their mission was clear: Secure the ammunition depot, protect the II Field Force command post, and eliminate the Viet Cong forces inside the perimeter.

The M113 armored personnel carriers proved invaluable. These tracked vehicles—essentially armored boxes on treads with

.50-caliber machine guns mounted on top—could absorb small arms fire while delivering overwhelming firepower. The thick armor protected infantry as they moved between positions. The heavy machine guns could suppress enemy positions that would have pinned down troops on foot.

Captain Moye, coordinating from the battalion headquarters element, watched his company maneuver through the ammunition depot complex. Buildings were on fire. Explosions from cooking off ammunition echoed constantly. Smoke obscured visibility. Viet Cong forces were mixed with American support personnel trying to defend their positions. It was chaos—but organized chaos. The battalion had trained for years. Officers knew how to coordinate multiple companies. Radiomen kept communications flowing. Platoon leaders led their men through the confusion.

Within hours, the 2/47th had secured the critical areas and eliminated or driven off the Viet Cong attackers. The ammunition depot was damaged but not destroyed. The command post was safe.

But there was no time to rest or celebrate. New orders came almost immediately: **Move to Saigon. Cholon district is under attack. Your battalion will clear it.**

Captain Moye looked at his maps. Cholon was Saigon's Chinatown—a dense, urban area with narrow streets, multi-story buildings, markets, shops, and tens of thousands of civilians. It was the opposite of the rural terrain where mechanized infantry normally operated. And they were expected to clear it of entrenched Viet Cong forces.

Starting now.

**This is the moment of forced adaptation—when your environment changes so completely and so quickly that you have no choice but to transform or fail.**

## INTO THE CITY

February 1–4, 1968. Cholon District, Saigon.

The 2/47th Infantry's M113s rumbled through Saigon's streets, an incongruous sight in an urban landscape. These vehicles were designed for off-road mobility, not navigating city blocks. They were loud, conspicuous, and limited in where they could maneuver.

But they had two advantages that proved decisive: armor and firepower.

Cholon's streets were narrow and congested. Buildings rose two, three, or sometimes four stories. Viet Cong forces had seized key structures and fortified them. Snipers fired from windows and rooftops. RPGs (rocket-propelled grenades) could be launched from alleys. Automatic weapons fire echoed from every direction.

Infantry on foot would have been shredded. But the M113s could absorb the small arms fire while moving forward. Their .50-caliber machine guns could suppress enemy positions. And critically, their armor protected the infantry dismounting behind them to clear buildings.

The tactics evolved in real time:

1. M113s would advance to a target building, providing suppressing fire
2. Infantry would dismount behind the vehicles, using them as cover
3. Engineers would breach the building—sometimes with explosives, sometimes just by ramming through walls
4. Infantry would clear room by room
5. Once secured, they'd move to the next building

Captain Moye coordinated this from the battalion tactical operations center (TOC), but he also moved forward to observe operations directly. He couldn't manage urban combat from a map alone—the situation changed too rapidly, and communication was frequently disrupted by buildings and the chaos of combat.

He described it later: "We trained for jungle. We trained for rice paddies. Nobody trained us for this. We figured it out as we went, lost some good men learning, and eventually cleared the district. But it was nothing like what we expected when we deployed."

**This is forced adaptation at its rawest: learning while doing, with lives on the line, with no time for formal training or careful planning.**

The tactical innovations the 2/47th developed in those four days weren't from a manual. They were improvised solutions to immediate problems:

- Using the M113's bulk to ram through street barricades
- Mounting extra machine guns for suppressive fire in urban environments
- Coordinating with helicopter gunships to strike upper floors while ground forces cleared lower floors
- Developing hand signals and communications protocols when radios were blocked by buildings
- Creating ad-hoc medical evacuation procedures in tight urban spaces

Some of these innovations worked. Some got people killed before they learned better. All of them were invented under pressure because the alternative was failure.

By February 4, the 2/47th and other U.S. and ARVN units had largely cleared Cholon. The Viet Cong had been killed, captured, or driven out. The district was secured. The cost: The 2/47th Infantry had eight men killed and dozens wounded in those four days. Viet Cong casualties were much higher—hundreds killed in Cholon alone—but those numbers provided little comfort to the units that had lost people learning urban warfare on the job.

**The lesson: Rapid adaptation is possible, but it's expensive. The organizations that survive disruption do so by learning faster than their competitors, but learning under pressure extracts a toll.**

## MINI-TET: MAY 1968

Tet wasn't a single event—it was a series of offensives that continued for months.

In May 1968, the communists launched what became known as "Mini-Tet"—a second wave of attacks on Saigon and other cities. The 2/47th Infantry, now experienced in urban combat, was again called to respond.

On May 7, 1968, they rushed from Bearcat to Saigon's Eighth District to reinforce U.S. military police and ARVN forces battling several Viet Cong battalions at Cau Mat hamlet and the Y-Bridge area.

Company B, 2/47th Infantry led the charge down a canal road toward Cau Mat. They were immediately hit with rockets and automatic weapons fire—a well-prepared ambush. Multiple soldiers were casualties in the opening salvos.

But this time, they knew what they were doing. The company commander didn't panic or retreat. He called for helicopter gunship support, coordinated with adjacent units, and pressed the attack using the combined arms tactics they'd developed during the February fighting.

The mechanized troops responded with overwhelming firepower—.50-caliber machine guns, 20mm cannons on some vehicles, coordinated with helicopter strikes. They expended so much ammunition that Lieutenant Colonel Tower, the battalion commander, had to arrange emergency resupply from Bearcat.

Through the night, Company B and the attached 6/31st Infantry fought house to house. By morning, they'd cleared Cau Mat, finding twelve dead Viet Cong and forcing the rest to retreat. The 2/47th had suffered eight killed—a terrible cost—but they'd accomplished the mission and prevented the Viet Cong from seizing a critical avenue into Saigon.

**This is what adaptation looks like after the initial shock: The second crisis is still difficult, still costly, but the learning from the first crisis makes you more effective.**

The 2/47th Infantry in May 1968 was a different organization than in January 1968. They hadn't had time for formal retraining. They hadn't gone through an official "urban warfare course." They'd simply been forced to adapt or fail, and they'd adapted—incorporating hard-won lessons into their operations, developing new tactics, and training new replacements in urban combat techniques that hadn't existed in Army doctrine.

**Organizations that survive disruption become different organizations—not through careful planning but through forced evolution under pressure.**

## THE LEADERSHIP CHALLENGE: LEARNING WHILE LEADING

Captain Moye faced a leadership challenge during Tet that mirrors what executives face during major market disruptions:

**How do you lead people through a crisis when you don't know the answers yourself?**

He wasn't an urban warfare expert. His training and experience were in rural counterinsurgency, civic action, and coordination with allied forces. Cholon required skills he hadn't developed. He couldn't say: "Follow my plan—I've done this before." He hadn't. He couldn't say: "Trust the doctrine—it will work." There was no doctrine for this situation. What he could do—what he had to do—was:

### *1. Acknowledge the reality*

He didn't pretend he had all the answers. In briefings with his officers, he was direct: "This isn't what we trained for. We're going to figure it out as we go. Some things will work. Some won't. Learn fast."

This honesty built trust. When leaders pretend to have expertise they don't have, teams lose confidence quickly. When leaders acknowledge uncertainty while demonstrating commitment to finding solutions, teams rally.

### *2. Empower subordinates to innovate*

The tactical innovations that made the 2/47th effective in Cholon didn't come from battalion headquarters. They came from platoon leaders and sergeants on the ground who were actually fighting. Captain Moye's role was creating the conditions where those innovations could emerge and spread:

- After action reviews where successful tactics were discussed
- Communication channels so different companies could learn from each other
- Authority for junior leaders to try new approaches without waiting for approval
- Recognition for soldiers who developed effective techniques

### *3. Protect the team's psychological safety*

Urban combat was terrifying in ways that rural operations weren't. Soldiers couldn't see the enemy until they were already engaged. Every building could hide threats. Casualties came suddenly and without warning. Captain Moye couldn't eliminate the danger, but he could ensure that people didn't feel abandoned or alone:

- Visiting forward positions personally, not just managing from the TOC
- Making sure medical evacuation was as fast as possible
- Ensuring that fallen soldiers were recovered and honored
- Maintaining the message: "We take care of our people; we complete our missions"

### 4. *Find and celebrate small wins*

In the chaos of Tet, it would have been easy to focus only on what went wrong—the casualties, the mistakes, and the situations where tactics failed. Captain Moye made sure to recognize what went right: the company that cleared a building without casualties, the platoon that developed an effective breach technique, and the soldiers who rescued wounded comrades under fire.

These small wins created momentum and confidence that sustained the organization through the crisis. **This is the leadership framework for navigating disruption: Acknowledge uncertainty, enable innovation, protect psychological safety, and recognize progress.**

## THE AI DISRUPTION PARALLEL

Fast-forward 55 years from Tet to November 2022. OpenAI released ChatGPT to the public on November 30, 2022. Within five days, it had one million users. Within two months, it had 100 million—the fastest-growing consumer application in history. And overnight, every company that touched information work faced their Tet Offensive moment.

The strategic landscape that had seemed stable—where Google dominated search, where content creation required human writers, where software development followed predictable patterns, and where customer service meant call centers—suddenly wasn't stable anymore. AI that could write coherent text, answer complex questions, generate code, analyze data, and simulate human conversation was available to anyone with an internet connection. For free.

Companies had the same basic choice Captain Moye faced in January 1968: Adapt immediately or become irrelevant.

Some adapted. Many didn't.

**Let me show you what adaptation—and failure to adapt—looks like in the AI era.**

## THE COMPANIES THAT ADAPTED

### *Microsoft*

In late 2022, Microsoft was widely perceived as a technology dinosaur—a company that had missed mobile, social media, and consumer internet. Their core products (Windows, Office, and Azure) were enterprise-focused legacy platforms.

Then CEO Satya Nadella made a decision that looked risky at the time: Invest $10 billion in OpenAI and commit to integrating AI throughout Microsoft's entire product line. This meant:

- Rebuilding Bing search with ChatGPT integration (directly challenging Google's monopoly)
- Integrating AI into Office 365 as Copilot (transforming how people use Word, Excel, and PowerPoint)
- Making Azure the preferred platform for AI development
- Essentially betting the company's future on AI-first products

Nadella described this as Microsoft's "platform shift moment"—comparable to the shift from mainframes to PCs, or from PCs to mobile. The stock market rewarded this adaptation: Microsoft's market cap increased by over $1 trillion in 2023, driven primarily by AI initiatives.

**This is the corporate equivalent of the 2/47th Infantry pivoting to urban combat—recognizing that your existing capabilities need to transform immediately, and committing fully to that transformation even though it's uncomfortable and risky.**

### *Canva*

Canva—the design software company that made graphic design accessible to non-designers—faced an existential threat from AI image generation tools such as Midjourney and DALL-E. Why would people use Canva's templates when they could just describe what they wanted and have AI generate it?

Their response was rapid and comprehensive:

- Integrated AI image generation directly into Canva
- Added AI writing assistants for marketing copy
- Built AI-powered design suggestions and layout optimization
- Positioned themselves as the AI-powered design platform for non-designers

They essentially said: "AI isn't replacing us—it's making our core value proposition (design for everyone) even more powerful."

By integrating AI rather than competing with it, Canva maintained relevance and continued growing even as standalone AI image generators gained traction.

**This is adaptation that preserves core purpose while transforming tactics—exactly what the 2/47th did by using their M113s in ways they'd never been designed for.**

### *Duolingo*

Duolingo, the language learning app, faced a potential disruption: AI chatbots that could provide free, unlimited conversation practice in any language. Why pay for Duolingo when you could just talk to ChatGPT in Spanish?

Their response:

- Built AI tutors that provide personalized conversation practice
- Created AI-powered pronunciation feedback
- Developed adaptive learning paths that adjust in real time based on AI analysis of user performance
- Maintained their gamification and engagement features that pure AI chatbots lacked

They recognized that AI wasn't their competitor—it was a tool that could make their product dramatically better. By incorporating it quickly, they strengthened their position.

**Companies that adapted successfully shared common characteristics:**

1. **Speed of recognition** – They understood immediately that AI was a fundamental shift, not a feature trend
2. **Willingness to cannibalize** – They were willing to disrupt their own products rather than wait for competitors to do it
3. **Full commitment** – They didn't just add "AI features"—they rebuilt around AI
4. **Preservation of core value** – They kept their fundamental purpose while transforming how they delivered it

## THE COMPANIES THAT FAILED TO ADAPT

Now let's look at the other side—companies that tried to iterate when they needed to pivot, that added AI features when they needed to rebuild, and that treated this a normal technology shift rather than a fundamental disruption.

### *Chegg*

Chegg was a homework help platform—essentially a tutoring service where students could get answers to homework questions. Their business model depended on students paying subscriptions for access to solutions and expert help.

Then ChatGPT launched. Students could get homework help for free, instantly, without waiting for a human tutor. Chegg's initial response was to add AI features—an AI tutor bot and AI-powered study guides. But they tried to keep their existing business model: subscriptions for access to AI-enhanced human help.

The problem: ChatGPT was already giving students what they needed, and it was free and immediate. Chegg's "AI-enhanced" offering was slower and more expensive than just using ChatGPT directly.

Result: Chegg's stock price collapsed from approximately $90 in 2021 to under $2 by late 2024—a 98 percent decline. They're now in survival mode, trying to rebuild completely.

**This is the equivalent of trying to win at Cholon using the same tactics that worked in the rice paddies—iterating on your existing approach when the environment demands transformation.**

### *Traditional Media Companies*

Many traditional media companies responded to AI by trying to protect their content rather than adapting their business models. The *New York Times* sued OpenAI for copyright infringement. Several publishers blocked AI crawlers from accessing their websites. Content creators demanded compensation for AI training on their work.

These aren't necessarily wrong positions legally or ethically. But strategically, they're defensive moves that don't address the fundamental disruption: AI is changing how people discover, consume, and interact with information.

Meanwhile, new AI-native media companies emerged:

- Perplexity AI (AI-powered search and research)
- Character.AI (conversational AI for entertainment and education)
- Numerous AI newsletter and content services

These companies weren't better journalists or better writers. But they understood the new distribution mechanism—AI as the interface layer between users and information—and built for that from the beginning.

**When your response to disruption is primarily defensive—protecting what you have rather than building what's needed—you're likely to lose.**

### *IBM Watson Health*

IBM invested billions in Watson Health, trying to use AI for medical diagnosis and treatment recommendations. They positioned it as "AI for healthcare" and sold it to hospitals and health systems. But Watson Health was built on older AI technology (pre-deep learning). When modern AI models emerged, Watson's approach looked obsolete. More importantly, IBM tried to sell Watson as a product that hospitals would buy and deploy, rather than build AI-native healthcare applications.

In 2022, IBM sold Watson Health for parts, effectively admitting the initiative had failed.

The missed opportunity: IBM had early-mover advantage in healthcare AI but failed to adapt when the technology paradigm shifted. They iterated on their existing approach instead of rebuilding for the new AI era.

## PATTERN OF FAILURE

Companies that failed to adapt shared characteristics:

1. **Denial about disruption speed** – Treating AI as an incremental improvement rather than a fundamental shift
2. **Defensive positioning** – Trying to protect existing business models rather than building new ones
3. **Surface-level AI integration** – Adding "AI features" without rebuilding core products
4. **Attachment to legacy revenue** – Unwilling to cannibalize existing revenue streams
5. **Slow decision-making** – Waiting for consensus and perfect information instead of moving quickly

## THE SPEED OF DISRUPTION

Here's what makes the AI disruption particularly challenging: the speed.

When Captain Moye faced the Tet Offensive, he had hours to adapt to urban combat—not weeks or months to retrain, but hours. When AI disruption hit, companies had similar timeframes. ChatGPT went from launch to 100 million users in two months. Companies that took six months to develop an AI strategy were already behind.

This speed differential is accelerating. Technology disruptions used to take decades:

- Electricity: ~40 years from invention to widespread adoption
- Telephone: ~30 years
- Personal computer: ~20 years
- Internet: ~15 years
- Mobile: ~10 years
- Social media: ~5 years
- AI: ~2 years from GPT-3 to mainstream adoption

**The time available to adapt is compressing.**

This creates specific leadership challenges:

### *Challenge 1: Decision Quality vs. Decision Speed*

In stable environments, you can gather information, analyze thoroughly, build consensus, and make careful decisions. In disruption, you have to make consequential decisions with incomplete information, quickly.

Captain Moye couldn't wait for perfect intelligence about Viet Cong positions in Cholon. He had to deploy his companies based on fragmentary reports and adjust in real time. Modern CEOs facing AI disruption can't wait for comprehensive market research about how AI will affect their industry. They have to make bets—invest in AI integration, restructure teams, and reallocate budgets—based on limited information.

**The leadership skill is making fast decisions that are "good enough" rather than slow decisions that are optimal.**

This requires:

- Comfort with uncertainty
- Willingness to decide and adjust rather than decide and commit
- Clear principles that guide decisions even when data is limited
- Trust in your team to execute and adapt

### *Challenge 2: Organizational Agility vs. Bureaucracy*

Large organizations optimize for stability, consistency, and risk management. They have approval processes, budget cycles, and strategic planning calendars. Disruption requires speed, experimentation, and rapid iteration. The approval processes that prevent mistakes also prevent quick adaptation.

The 2/47th Infantry couldn't convene a committee to review urban combat tactics and make recommendations for the next budget cycle. They had to try things, see what worked, and immediately apply lessons learned. Companies facing AI disruption need similar agility. But most are structured for efficiency in stable markets, not adaptation in volatile ones.

**The leadership challenge is creating "fast lanes" within bureaucratic organizations—ways to move quickly on critical adaptations without dismantling all the systems that prevent chaos.**

Some companies have done this successfully:

- Amazon's "two-pizza teams" that can make decisions without committee approval
- Google's "20% time" (before they killed it) that allowed engineers to experiment
- Microsoft's AI-specific budget pool with streamlined approval for AI investments

### *Challenge 3: Skill Evolution vs. Workforce Inertia*

The 2/47th Infantry's biggest constraint during Tet wasn't equipment—it was that their soldiers knew rural tactics but not urban ones. They had to learn on the job. Companies facing AI disruption have the same problem: Their workforce has skills optimized for the pre-AI world. Retraining takes time. Hiring new people with AI skills is competitive and expensive. Waiting means falling further behind.

**The leadership challenge is accelerating skill development while maintaining current operations.**

Microsoft did this by:

- Massive internal AI training programs for existing employees
- Aggressive hiring of AI talent
- Partnering with OpenAI to get access to cutting-edge capabilities their own teams couldn't yet build
- Creating AI-specific career paths to retain AI-skilled employees

But this works only if leadership commits fully and moves fast. Gradual training programs on a three-year timeline are too slow when disruption happens in months.

## THE AI BUBBLE QUESTION

Is there an AI bubble? Will it burst?

Almost certainly yes to both. Here's why: When a new technology emerges, investment floods in. Everyone wants to be part of the next big thing. Capital becomes cheap for anything with "AI" in the pitch deck. Valuations become disconnected from fundamentals.

This creates excess: Companies that shouldn't get funded get funded, products that aren't ready get shipped, and promises that can't be kept get made. Eventually, reality asserts itself. Some companies fail. Some products disappoint. Some promises prove empty. Investment contracts. Valuations collapse.

**This is a bubble, and it will burst.**

But—and this is crucial—**the bursting of the bubble doesn't mean the technology was overhyped or won't transform everything.**

The internet bubble of 1999–2000 was absolutely a bubble. Pets.com, Webvan, eToys—hundreds of companies with absurd valuations collapsed. The NASDAQ lost 78 percent of its value between 2000 and 2002.

But the internet still transformed everything. Amazon, Google, Facebook, Netflix—the companies that survived the bubble became some of the most valuable in history. E-commerce, cloud computing, social media, streaming—all the innovations that were promised during the bubble eventually happened, just not as fast or with the same companies that initially promised them.

**The AI bubble will follow the same pattern:**

Phase 1 (2023–2024): Exuberance – Everyone builds AI features, every startup claims to be "AI-powered," investment floods in, and valuations soar.

Phase 2 (2024–2026?): Disillusionment – Many AI features don't work well, some startups fail to deliver, returns don't justify investment, and the bubble bursts.

Phase 3 (2026+): Actual Transformation – The survivors figure out real use cases, technology matures, real value gets created, and AI actually transforms industries.

**For leaders, this creates a specific challenge: How do you adapt quickly enough to survive the disruption, without overcommitting to approaches that won't survive the bubble bursting?**

The answer is the same framework we've been discussing:

1. **Be in love with the problem, not the solution** – Commit to using AI where it solves real problems, not using AI for AI's sake.
2. **Iterate and adapt** – Build, test, learn, and adjust. Don't commit huge resources to untested AI bets.
3. **Preserve your force** – Don't bet the company on AI positions that might not work. Keep cash reserves, maintain core business, and stay flexible.
4. **Learn faster than competitors** – The companies that win won't be those that made the biggest AI bets earliest—they'll be the ones that learned fastest what actually works.

This is exactly what Captain Moye did during Tet: Tried tactics, saw what worked, adjusted immediately, and preserved his force for the longer campaign.

## THE TACTICS THAT WORK: LESSONS FROM TET AND AI

Looking across both the Tet Offensive and the AI disruption, certain tactical approaches consistently separate successful adaptation from failure:

### *Tactic 1: Parallel Development*

During Tet, the 2/47th Infantry didn't stop all other operations to retrain for urban combat. They adapted in real time while continuing their mission. Similarly, successful AI adaptation doesn't mean shutting down your existing business to rebuild completely. It means running parallel tracks:

- Maintain and optimize your current business
- Simultaneously build AI-native capabilities
- Figure out the transition from current to future

Microsoft didn't shut down Office to rebuild it AI-first. They maintained Office while building Copilot in parallel, and then integrated them.

### *Tactic 2: Modular Architecture*

The M113 armored personnel carriers proved valuable in urban combat because they were modular—soldiers could mount different weapons, reconfigure the interior, and adapt them to new purposes.

Companies with modular architectures (microservices, APIs, or cloud-native design) can integrate AI more easily than companies with monolithic legacy systems. This is why cloud companies adapted to AI faster than on-premise companies—their architectures were already designed for rapid change.

### *Tactic 3: Decentralized Experimentation*

The urban combat innovations during Tet came from platoon leaders and sergeants on the ground, not from battalion headquarters. Central command created the conditions for experimentation, but the innovations emerged from the people closest to the problem.

Similarly, successful AI adaptation often comes from teams close to customers—product managers, engineers, and customer success people—who see specific problems AI could solve. Smart companies create internal AI funds or "innovation time" where teams can experiment with AI applications without needing lengthy approval processes.

### *Tactic 4: Rapid Information Sharing*

The 2/47th's advantage over Viet Cong forces wasn't individual soldier skill—it was organizational learning. When one company discovered an effective tactic, other companies learned it immediately through after-action reviews and radio communications.

Similarly, companies that adapt fastest have mechanisms for sharing learnings across teams:

- Internal AI showcases where teams demonstrate what they've built
- Documentation of what works and what doesn't
- Communities of practice where AI experimenters share lessons

### *Tactic 5: Ruthless Prioritization*

During Tet, Captain Moye couldn't solve every problem. He had to prioritize: Secure Long Binh first, then clear Cholon, and then handle other threats. In AI disruption, companies can't rebuild everything at once. They need to identify:

- Which functions AI will disrupt fastest (do these first)
- Which capabilities create competitive advantage (invest heavily)
- Which are necessary but not differentiating (use vendor solutions)
- Which can wait (deprioritize)

Many companies fail by trying to "AI-ify" everything simultaneously, spreading resources too thin to be effective anywhere.

## THE HUMAN ELEMENT

There's one final parallel between Tet and AI disruption worth exploring: the human psychological challenge.

Captain Moye's soldiers during Tet weren't just learning new tactics. They were also processing fear, confusion, and the psychological shock of discovering that their entire understanding of their mission had been wrong. They thought they were winning a rural pacification campaign. Tet proved they weren't winning anything—the enemy could strike anywhere, anytime.

This cognitive dissonance was debilitating for many. Some soldiers broke psychologically. Some became cynical. Some just checked out mentally while going through the physical motions. Captain Moye's leadership during this period wasn't just tactical—it was psychological. He had to help his men process the shock, find new meaning in their mission, and maintain commitment despite the strategic confusion.

**AI disruption creates similar psychological challenges for employees.** The workers who spent years becoming expert at pre-AI skills—writers, programmers, analysts, and customer service representatives—are watching AI potentially make their expertise obsolete. This creates:

- Fear about job security
- Resentment toward leadership for not protecting them
- Cynicism about company AI initiatives
- Identity crisis about professional worth

Smart leaders address this directly:

### *1. Acknowledge the disruption honestly*

Don't pretend AI is just another tool that will enhance everyone's job. For some people, it will replace their roles. Being honest about this—while committing to help people transition—builds trust. Captain Moye didn't tell his troops, "Urban combat is just like what we've been doing." He said, "This is different; we'll figure it out together."

### *2. Provide clear paths forward*

People can handle disruption if they see a path to remaining valuable. Companies should:

- Offer training in AI-adjacent skills
- Create new roles that combine domain expertise with AI capabilities
- Show how existing skills remain valuable in AI-augmented workflows

Microsoft told their employees: "AI won't replace you, but people who know how to use AI effectively will replace people who don't. We'll teach you."

### *3. Recognize loss*

When skills become obsolete, people experience grief. Leaders should acknowledge this rather than toxic positivity ("This is great! AI makes everyone more productive!").

Better: "I know this is hard. Skills you spent years developing are changing. We'll support you through this transition."

### *4. Maintain human connection*

During Tet, Captain Moye visited forward positions, ate with his soldiers, and attended memorials for the fallen. He maintained human connection, even in chaos. During AI disruption, leaders need to do the same—maintain face time, listen to fears, and demonstrate that people matter beyond their economic productivity.

## *From Cholon to the Corporate Boardroom*

February 4, 1968. After four days of fighting, the 2/47th Infantry had cleared Cholon. The district was secured. The Viet Cong had been defeated. Captain Moye walked through the streets his battalion had fought through. Buildings were damaged by gunfire and explosives. Debris littered the roads. Civilians were beginning to emerge from hiding. He thought about the men he'd lost—eight killed during the February fighting, dozens wounded. He thought about the tactical innovations his soldiers had developed under pressure. He thought about how completely different this operation was from anything they'd trained for.

And he understood something fundamental about warfare—and leadership—that would stay with him for the rest of his life:

**The plan never survives contact with the enemy. Adaptation isn't optional. The organizations that survive are the ones that learn fastest, not the ones that planned best.**

Fifty-five years later, in corporate boardrooms and startup offices, leaders are learning the same lesson. The companies that survive the AI disruption won't be the ones that had the best AI strategy in 2023. They'll be the ones that adapted fastest when their 2023 strategies proved insufficient. They'll be the ones that:

- Recognized the disruption immediately rather than dismissing it
- Moved quickly even with incomplete information
- Empowered their teams to experiment and innovate
- Preserved resources and flexibility rather than betting everything on one approach
- Learned from failures and adjusted rapidly
- Maintained organizational cohesion through the transition

**This is the leadership lesson from Cholon: When your environment changes completely overnight, survival depends on adaptive intelligence—the ability to learn, adjust, and execute under pressure.**

The 2/47th Infantry didn't win at Cholon because they were better trained or better equipped than the Viet Cong. They won because they adapted faster, learned from each engagement, and maintained organizational effectiveness through chaos.

The companies that win in the AI era won't win because they had better AI technology in 2023. They'll win because they adapted faster, learned from each iteration, and maintained organizational effectiveness through disruption.

The tactics differ. The technology differs. The stakes differ. But the fundamental challenge is identical: How do you lead people through a transformation you didn't expect, using skills you don't yet have, to fight battles you weren't trained for?

Captain James Moye III answered that question in the streets of Saigon in 1968. The question is: How will you answer it now?

# CHAPTER 7

## *Red Noses in Norway*

June 2003. Andøya Air Station, Norway.

WE SAT IN THE tiny break room at Andøya—literally at the top of the world, 300 kilometers north of the Arctic Circle—watching the sun refuse to set. It was 11:30 PM, and daylight still flooded through the windows. In early June this far north, the sun barely dips below the horizon before coming back up. It's disorienting, like living in a movie where someone forgot to add the night scenes.

Combat Air Crew Four had arrived in Norway a week earlier, exhausted from our Iraq deployment but with no time to decompress. The Navy needed submarine hunters, and we were still one of the few P-3 crews fully qualified for anti-submarine warfare operations. We had received intel that a Russian sub has left port in Polyarny and was headed for the Atlantic.

We'd been sitting on "ready status" for four days—meaning we could get called to launch at any moment. Our gear was staged. The aircraft was preflighted and ready. We lived at the air station, waiting.

This is the nature of submarine hunting: You can't schedule it. You can't say, "We'll hunt submarines Tuesday afternoon from 2–5 PM." You hunt when intelligence indicates a contact, when the submarine is where you can reach it, and when weather and conditions align.

You wait. And then when the call comes, you launch immediately and hope you can find your target.

At 0130 hours on June 8, the phone rang in the ready room. Our operations officer listened, hung up, and turned to us: "We've got a contact. Akula-class Russian submarine, northbound through the Norwegian Sea. Brief in thirty minutes, launch at 0300."

We were going hunting.

## THE SUBMARINE HUNTER

Before I explain what happened during that mission, you need to understand what the P-3C Orion is and what submarine hunting actually involves. The P-3C Orion is a four-engine turboprop aircraft that first flew in 1962. It was designed specifically to hunt Soviet submarines during the Cold War—to patrol the vast ocean expanses where Russian ballistic missile submarines operated, to track them, and to know exactly where they were at all times.

This was a critical strategic capability during the Cold War. The Soviet Union had dozens of submarines carrying nuclear missiles. If those submarines could hide effectively, they'd constitute an existential threat—a second-strike capability that could destroy American cities, even if a first strike eliminated Russia's land-based missiles.

So the U.S. Navy spent billions developing systems to find and track submarines. The P-3 was the airborne component of that system. Here's how it works:

**Sonobuoys**: These are expendable acoustic sensors—essentially hydrophones in small cylindrical packages. You drop them from the aircraft into the water. They deploy a hydrophone on a cable that dangles at a specific depth, listening for submarine sounds. They transmit what they hear back to the aircraft via radio.

A P-3 carries approximately 48 sonobuoys in pre-loaded launch tubes in the belly of the aircraft, plus another 40–60 buoys carried inside that were launched out of three chutes operated by a technician in the back of the aircraft (but we could carry more if needed). Different types listen at different depths and different frequencies for different purposes. Some are passive (just listening). Some are active (they emit a sound pulse and listen for reflections, like underwater sonar).

**Acoustic Analysis**: Inside the P-3, we have sensor operators—enlisted specialists who analyze the acoustic data from the sonobuoys. They wear headphones and watch screens showing frequency analysis of underwater sounds.

Experienced sensor operators can distinguish between:

- A whale (biological sounds have distinctive patterns)
- A merchant ship (propeller cavitation has specific acoustic signatures)
- A Russian submarine (each class of submarine has unique acoustic characteristics—machinery noise, propeller design, and hull resonance)

They can tell you not just that it's a submarine, but specifically that it's an Akula-class attack submarine versus a Kilo-class diesel boat versus an older Victor-class.

**Magnetic Anomaly Detection (MAD)**: The P-3 has a long boom extending from its tail with a magnetometer. Submarines are large metal objects. When you fly directly over one, the magnetometer detects the distortion in Earth's magnetic field caused by all that metal.

MAD works only if you're very close—you have to fly within a couple thousand feet horizontally and very low (we typically fly 200–500 feet above the water for MAD operations). But when you get a MAD contact, you know precisely where the submarine is.

**Radar**: For surface contacts and navigation. It's not useful for finding submerged submarines, but it is helpful for avoiding other ships and aircraft unless a submarine comes to the surface or even deploys its periscope.

**The Coordination Challenge**: All of this requires a crew working together with extraordinary precision:

- Pilots fly the planned search pattern, maintain altitude and airspeed, and respond to tactical coordinator directions (called " Flight").
- The tactical coordinator (TACCO—that was my role) manages the overall mission, decides where to drop sonobuoys, directs the search pattern, and coordinates with other assets.
- The navigator plots positions, calculates intercepts, and maintains the tactical picture.
- Sensor operators listen to sonobuoys, analyze contacts, and classify submarine signatures. The two sensor operators who listened to the buoys were called "Jezebel" or just "Jez." The nickname "Jezebel" for Navy acoustic sensor operators stems from Project Jezebel, the Cold War codename for the Navy's long-range, passive acoustic surveillance system (SOSUS). The other sensor operator who managed the MAD boom and radar was sensor 3, or just "three" on the radio.
- In-flight technicians (IFTs) load and deploy sonobuoys and operate the MAD system.
- Flight engineers manage aircraft systems, fuel consumption, and engine optimization.

It's the ultimate team sport. No single person can hunt a submarine. The TACCO can't hear the acoustic data. The sensor operators can't see the tactical picture. The pilots can't analyze contacts while flying. Everyone depends on everyone else.

**And here's the critical thing: You're hunting something that's actively trying not to be found.**

Modern submarines are extraordinarily quiet. Russian Akula-class submarines are nearly as quiet as American Los Angeles–class attack subs—which was shocking when they first appeared in the 1980s. They have anechoic coatings that absorb sonar pings. They can operate at multiple depths. They can go silent by shutting down machinery and drifting.

The submarine crew knows you're looking for it. The submarine commander is actively trying to evade detection—changing depth, changing course, and minimizing noise.

**This is the ultimate test of adaptive intelligence: hunting something intelligent that's hunting you back (in exercises) or evading you (in real operations), where you can't see it, where it can hide in an environment you can't fully map, and where every decision you make gives it information about where you are and what you're doing.**

## THE HUNT BEGINS

0300 hours. We launched from Andøya into the pale Arctic twilight.

Our intelligence briefing had given us an initial datum—a location where the Russian Akula had been detected by Norwegian coastal sensors. The submarine was northbound, probably heading back to its base at Polyarny on the Kola Peninsula after whatever mission it had been conducting.

My tactical picture showed:

- Last known submarine position: approximately 200 nautical miles northwest of Andøya

- Submarine speed: estimated 12–15 knots based on previous track
- Our transit time: approximately 60 minutes to reach the search area
- Water depth: 1,200–1,800 meters (deep water, lots of room for the submarine to maneuver)
- Weather: broken clouds, light winds, decent conditions for acoustic operations

The standard approach to submarine hunting is to lay a trap—what we call a "barrier search." You drop a field of sonobuoys in a pattern that creates a detection barrier across the submarine's expected path. Then you wait and listen. When the submarine passes through your barrier, the sonobuoys detect it, and you can begin actively tracking.

But here's the problem: A submarine can be anywhere in three dimensions. The ocean is vast. Sonobuoys have limited range (typically 1–2 nautical miles detection radius, depending on conditions and submarine noise level). And we had only 87 sonobuoys onboard.

If you spread them too far apart, the submarine can slip through the gaps. If you concentrate them in one area, the submarine might be somewhere else entirely. If you put them at the wrong depth, the submarine might be above or below the thermal layer where your sonobuoys can hear it.

**This is the fundamental challenge of submarine hunting: You're making probabilistic bets with limited resources about where an adversary will be, while that adversary is actively trying to foil your strategy.**

As we flew northwest toward the search area, I worked with my navigator and sensor operators to design our initial search pattern:

"Okay, based on his last position and speed, he should be in this box [marking an area on the tactical display]. We'll lay a 24-buoy barrier pattern across his probable track, with buoys at 400-foot depth. Water temperature shows a thermal layer at 300 feet, so 400 should put us in a good acoustic channel."

A note here on oceanography. Not only did we have to understand Russian submarine tactics, but we also had to have an in-depth understanding of the way sound waves propagate in the water and how temperature and pressure impact it. One of the easiest ways for a submarine to avoid being detected is to move in and out of the "layer." The layer is where warmer, shallower water heated by the sun meets the cold, frigid pressurized water of the deep ocean. Sound waves tend to get trapped on either side of this boundary, so if you are listening on one side or the other, and the sub moved out of where you were listening, you would lose contact. This depth varied depending on where in the world you were, the time of year, and even time of day. In the Norwegian Sea in the early summer, the layer was pretty shallow, 300 feet or so.

My senior sensor operator, a grizzled and experienced first-class petty officer that we called "Billy the Kid"—because he was grizzled and had been hunting submarines for twenty years—raised a concern: "TACCO, if he knows we're looking for him, he might go shallow. Above the layer. Our 400-foot buoys won't hear him if he's at 100 feet."

He was right. If the submarine commander knew he was being hunted and went shallow, our entire search pattern would be useless. But we had limited sonobuoys. We couldn't cover multiple depth bands and multiple geographic areas simultaneously.

**This is decision-making under uncertainty: imperfect information, resource constraints, an intelligent adversary, and consequences for being wrong.**

"We'll put three buoys shallow—90-foot depth—at key choke points," I decided. "If he's shallow, we might get lucky with one of those. But I'm betting he's at moderate depth, trying to move fast to get out to the Atlantic. Let's go with the 400-foot pattern."

It was a guess—an educated guess based on training, experience, and tactical analysis, but ultimately a guess. We arrived at the search area. I directed the pilots to fly a specific course and speed. At precise intervals, my in-flight technician deployed sonobuoys—small cylinders dropping from tubes in the belly of the aircraft, hitting the water with small splashes, and deploying their hydrophone cables.

On our tactical display, each sonobuoy appeared as a symbol showing its position and status. Within fifteen minutes, we had 24 sonobuoys in the water, creating a barrier across the submarine's expected path. Then we waited. The sensor operators put on their headphones and began monitoring each sonobuoy channel. They heard ocean ambient noise (which sounds like static), biologics (whales and fish), and distant merchant ship traffic. No submarine.

We flew a pattern at low altitude designed to keep us within radio range of all our sonobuoys while giving us reaction time if we detected the submarine.

Thirty minutes passed. Nothing. Sixty minutes. Still nothing.

Our fuel gauge showed we'd burned through about 25 percent of our on-station gas. We had maybe four more hours before we'd have to head home. Bernie, our other sensor operator, called on the intercom: "TACCO, I'm hearing zilch. Either he's not in this area, or he's too quiet for us to detect, or he's shallower than our buoys can hear."

**This is the moment of adaptive intelligence: Your plan isn't working. Do you persist? Do you adjust? Do you abandon the approach entirely?**

## REAL-TIME ADAPTATION

I looked at my tactical display. We had intelligence saying the submarine should be here. We had a carefully designed search pattern. We had experienced operators listening. And we had nothing. Two choices:

**Option 1: Persist with the current plan.** Maybe the submarine was delayed. Maybe our speed estimate was wrong and he hadn't reached our barrier yet. Maybe he'd be along any minute. If we repositioned and he passed through right after we left, we'd miss him entirely.

**Option 2: Adapt based on new information (or lack thereof).** Our assumption was wrong. Either he's not on this track, or he's at a different depth, or he took evasive action. Sitting here longer won't help. We need to try something else.

This is the challenge of submarine hunting—and of any situation where you're searching for something in uncertain conditions with limited information: **Persistence looks stupid when you're searching in the wrong place, but it looks smart when you just need to be patient. Adapting looks smart when your initial approach is wrong, but it looks stupid if you were close to success and just needed to wait.**

You can't know which is true until afterward. I made a call: "We're adapting. Billy, based on water conditions and the lack of contacts, where else could he be?"

Billy thought for a moment. "If he's shallow—above the thermal layer—he could have passed through already and we never heard him. Or if he changed course west, he'd be outside our barrier. Or..." he hesitated and then continued: "or maybe he's not northbound at all. Maybe he's sitting quiet, waiting for us to waste our sonobuoys and gas searching for him."

All plausible. All possible. But I had to choose one. I felt that the intel we had was good and that the sub was in the area where we thought, but its crew had heard us deploy our pattern and was waiting us out. This is the great game of chess that submarine commanders and P-3 TACCOs play—what move do you make next? Russian sub commanders were known to go full quiet, meaning shut everything down and wait out the hunter. I made a call.

"Flight, let's climb. Make it really noisy. Let him know we have expended our gas and are out of here. Billy, keep listening to the buoys." The gambit here was to make a show of leaving the area and gaining altitude so the sub could no longer hear us overhead. The idea was maybe the sub commander would buy our deception, think we left, and start up his engines again. It was a ploy, but I had nothing to lose.

The aircraft roared over the water, and we climbed hard to about 10,000 feet, much higher than the sub could hear us. In the meantime, we continued to wait and listen. Our head pilot came back to talk to me after 30 minutes.

"Hey, you sure about this?" he asked.

"Nope," I answered. "But we had to try something."

Within 10 seconds, Billy called over the intercom:

"TACCO, I have a contact! Sonobuoy 27, bearing 315, designate Contact 1."

My heart rate jumped. "Classify."

Bernie's voice was calm and professional: "Low-frequency machinery noise, consistent with Russian nuclear submarine. Turbine frequency matches Akula-class. High confidence this is our guy."

**There he was.** Not exactly where our initial intelligence said he should be. Not at the depth we'd originally assumed. But there—because we'd adapted when our first approach didn't work.

## THE GAME BEGINS

There is a saying in the P-3 community that there's nothing more fun with your clothes on than hunting submarines, and it's true. But once you detect a submarine, the real work begins. Detection is just the start—you need to track it, maintain contact, and gather as much information as possible about its capabilities and tactics. The submarine crew, meanwhile, tries to evade. They know it's been detected (submarines can identify sonobuoys deploying, and they can often detect aircraft overhead). So now it's a game: We try to follow it, and the crew tries to lose us.

"I've got bearing drift," Billy reported. "Contact 1 is moving. Bearing now 320, bearing rate indicates he's heading northwest, speed approximately 10 knots."

I quickly calculated: At 10 knots, the commander would move the submarine through our sonobuoy field in about 20 minutes. We needed to stay ahead of him—keep dropping sonobuoys in his path so we would maintain continuous contact. "Drop pattern P-2," I directed. "Four buoys around his track."

The TACCO can fire buoys from the belly of the plane, and the in-flight technician (IFT) can load buoys from the interior of the aircraft. We deployed the pattern, and the buoys splashed into the water, deployed their sensors, and reported back. For the next two hours, we played this game: The submarine maneuvered, changed depth, and varied its speed. We repositioned, dropped new sonobuoys, and adjusted our pattern based on its movements.

Every decision was probabilistic:

- It's heading northwest now, but will it continue or turn?
- It's at 700 feet now, but will it go deeper or shallower?
- It's making 10 knots now, but will the commander speed it up to try to outrun us or slow it down to quiet its signature?

We weren't reacting to certainty. We were anticipating, predicting, and placing bets with our limited sonobuoys about where we thought the sub would be in fifteen minutes. Sometimes we were right—we'd drop sonobuoys ahead of its track and pick it up right where we expected. Sometimes we were wrong—it zigged when we expected it to zag, and we'd have a gap in coverage.

**This is adaptive intelligence in its purest form: making rapid decisions with incomplete information, constantly adjusting based on new data, and learning from each iteration what works and what doesn't.**

At one point, we lost it. The last sonobuoy that had contact went out of range, and the next buoys we'd dropped showed nothing. "TACCO, I've lost Contact 1," Curtis reported. "No contact on any buoys."

Damn. The submarine commander either changed depth, changed course more dramatically than we'd anticipated, or went ultra-quiet and drifted below our detection threshold. I had maybe 30 seconds to decide: Where do we search next? I looked at the tactical picture—his last known position, his last known heading, the bathymetry, and my remaining sonobuoy count (we were down to about 35).

He'd been heading northwest. If he continued northwest, he'd run into shallower water—less depth to hide in. So maybe he turned north or northeast to stay in deep water. Or maybe that's what he wanted me to think, and he actually continued northwest, betting I'd search east.

**This is what it's like to compete against an intelligent adversary: second-order thinking where you're not just reacting to what they did, but also anticipating what they think you think they'll do.**

"Lost contact procedures," I called. "Eight buoys spread to cover both northeast and northwest options."

We dropped the buoys. Five minutes of silence—nothing. Then: "Contact! Sonobuoy 41, Contact 1 reacquired, bearing 005, he turned north like you predicted."

Relief. We had him again.

This continued for another hour—tracking, losing contact, reacquiring, and adapting our pattern based on his maneuvers. It was mentally exhausting, like playing high-speed chess where every move costs you a piece you can't get back (sonobuoys are expendable—once you drop them, they're gone) and your opponent is making moves you can't see directly.

Finally, our fuel state reached the point where we had to head home. We'd tracked the submarine for over three hours total, gathered acoustic data on its signature, documented its tactics, and maintained contact through multiple evasion attempts. We turned over our contact to another NATO asset who then picked up and carried on.

Mission successful.

## RED NOSES

When we landed back at Andøya, our maintenance chief met us at the aircraft with a red Sharpie.

"First time tracking a Russian sub for your crew?" he asked.

"First time for most of us," I confirmed. A few of our crew had hunted submarines before on previous deployments, but for most—including me as TACCO—this was our first successful tracking of a real Russian submarine.

The maintenance chief grinned. "Well, congratulations. You know the tradition."

He proceeded to pass around the red Sharpie, and we gave ourselves red noses. In P-3 squadrons, red noses signified that the crew had successfully tracked its first Russian submarine. It's called earning your "red nose"—a tradition that goes back to the

Cold War, when tracking Soviet submarines was the primary mission, and doing it successfully was both difficult and strategically important.

We took pictures of our crew with our red noses in front of our aircraft and the mountains of Norway. The tradition was silly—red Sharpie marking our noses—but it meant something. It meant we'd done what we'd trained for years to do. It meant we'd succeeded at one of the hardest challenges in naval aviation. But more than that, it represented what I now understand to be the core lesson of submarine hunting—and of adaptive intelligence more broadly:

**No amount of planning can substitute for doing. You learn by engaging with reality, adapting in real time, and iterating based on feedback.**

We could have spent another month in classroom training, studying submarine tactics, and planning perfect search patterns. But we learned more in four hours of actually hunting a real submarine than we would have in any amount of simulated training, because the real submarine did things our training scenarios hadn't anticipated. The real ocean had acoustic conditions that differed from our predictions. The real mission required decisions that couldn't be scripted in advance.

**The only way for us to learn was to do it.**

## NOTHING BEATS ACTION

This is the leadership lesson I took from that mission—and from submarine hunting generally: **In uncertain, complex, competitive environments, action beats planning.**

This isn't an argument against planning. We'd spent years training for submarine hunting. We had tactical publications, standard search patterns, and classification guides. Planning and preparation mattered. But at a certain point, more planning doesn't improve outcomes. At a certain point, the only way to learn is to engage with reality and adapt based on what you discover.

I see this pattern constantly in business, particularly in startups and innovation projects:

### *Failure Mode: Over-Planning*

Team A wants to build a new product. Before they start, they:

- Conduct extensive market research (6 months)
- Interview dozens of potential customers (3 months)
- Develop detailed specifications (2 months)
- Create comprehensive project plans (1 month)
- Design the perfect architecture (2 months)
- Finally start building (14 months after the initial idea)

By the time they launch, the market has changed, their assumptions are outdated, and their carefully planned product doesn't match what customers actually need now.

### *Success Mode: Iterative Action*

Team B wants to build a new product. They:

- Talk to a handful of customers about the problem (2 weeks)
- Build a minimal prototype that addresses the core problem (4 weeks, or 4 days now with AI)
- Put it in front of real users and watch what happens (1 week)
- Learn what works and what doesn't based on actual usage (1 week)
- Rebuild based on real feedback (4 weeks)
- Repeat the cycle (ongoing)

By the time Team A launches its carefully planned product, Team B is on its fifth iteration, each one informed by real user feedback and real market conditions.

**This is the "red noses in Norway" principle: You learn more from four hours of actual hunting than from four months of planning hunts.**

## THE STARTUP APPLICATION

The tech startup world has a term for this: "shipping and learning" or "build-measure-learn."

Eric Ries popularized this in *The Lean Startup*: instead of spending years building a perfect product, you build a minimum viable product (MVP), release it to real users, measure what happens, learn from the data, and iterate.

But most companies get this wrong. They say they're doing "lean startup" or "agile development," but they're actually just doing waterfall development in shorter cycles.

Real shipping-and-learning looks like submarine hunting:

### *1. Start with a hypothesis*

Before we launched from Andøya, we had a hypothesis: "The submarine will be in this location, at this depth, heading in this direction."

Before launching a product, you should have a hypothesis: "Users need X solution to problem Y, and they'll engage with it in Z way."

### *2. Test the hypothesis with real engagement*

We didn't simulate the submarine hunt. We actually went out and looked for a real submarine with real sonobuoys.

You can't simulate market engagement. You have to put a real product in front of real users and see what actually happens.

### *3. Gather data*

Our sensor operators listened to sonobuoys, analyzed acoustic data, and classified contacts. This gave us real-time feedback about whether our hypothesis was correct.

You need instrumentation—analytics, usage tracking, and user feedback mechanisms—that gives you real-time feedback about whether users are engaging how you expected.

### *4. Adapt immediately based on what you learn*

When our first search pattern found nothing, we didn't persist indefinitely. We adapted—moved our location, changed the depth, and adjusted our approach.

When users aren't engaging with your product as expected, you adapt—change features, adjust messaging, and pivot to what's actually working.

### *5. Iterate rapidly*

We didn't drop all 87 sonobuoys in one pattern and then wait. We dropped 24, evaluated, changed tactics, evaluated, dropped 4 more, and evaluated again.

You don't build your entire product roadmap at once. You build one feature, evaluate, build the next based on what you learned, evaluate, and repeat.

**The key insight: Each iteration is faster and more informed than the last, because you're learning from reality rather than guessing.**

## THE AIRBNB EXAMPLE

Airbnb is the canonical example of shipping-and-learning done right.

In 2008, Brian Chesky and Joe Gebbia had an idea: Let people rent out their spare rooms to travelers. They built a simple website, listed their own apartment, and tried to get users.

It didn't work. Almost nobody booked. They could have spent six months doing market research about why it didn't work. Instead, they looked at their data: the listings that did get bookings had good photos, while listings with bad photos (blurry phone pictures, poor lighting) got almost no bookings.

Hypothesis: Professional photography would increase bookings.

Test: They rented a camera, went to New York, and took professional photos of listings themselves.

Result: Those listings got two to three times more bookings.

Learning: Photo quality matters enormously for this product.

Action: They built a program where Airbnb would send professional photographers to photograph listings for free.

This entire cycle—hypothesis, test, result, learning, and action—took weeks, not months. And it was based on real data from real users, not surveys or focus groups.

They did this repeatedly:

- Hypothesis: Users don't trust staying in strangers' homes → Test: Add verified ID and reviews → Result: Bookings increased → Action: Invest in trust and safety
- Hypothesis: Hosts don't know how to price → Test: Suggested pricing algorithm → Result: More bookings at optimal prices → Action: Build sophisticated dynamic pricing
- Hypothesis: Users want unique experiences → Test: Add "Experiences" category → Result: Strong engagement → Action: Expand experiences globally

**Each iteration was informed by real usage data from the previous iteration.**

By contrast, there were dozens of "Airbnb but better" startups that launched with more features, better design, and more careful planning—and all failed, because they were planning in a vacuum rather than learning from reality.

**This is the difference between submarine hunting and pretending to hunt submarines: Real feedback creates real learning.**

## THE DATA IMPERATIVE

There's a critical requirement for this approach: **You need data.**

During our submarine hunt, we had sonobuoys giving us acoustic feedback. We had bearings and frequencies. We had classification algorithms. We had sensors that told us what was actually happening in the ocean. Without that data, we'd be flying blind—literally just guessing where the submarine might be with no feedback about whether our guesses were correct.

The same is true in product development. You can't "ship and learn" if you're not actually learning—and learning requires measurement. Too many companies "launch" products without proper instrumentation:

- No analytics tracking how users actually use features
- No cohort analysis showing which user segments engage vs. churn
- No A/B testing to compare different approaches
- No user feedback mechanisms to understand why they behave as they do

This is like hunting submarines without sonobuoys. You can drop search patterns all you want, but you'll never know if you found the submarine or not.

**The modern equivalent of sonobuoys is your analytics stack:**

- Event tracking (What actions do users take?)
- Funnel analysis (Where do users drop off?)
- Cohort retention (Do users come back?)
- Feature usage (What do users actually use?)
- User feedback (Why do they behave this way?)

Companies that build this instrumentation can adapt rapidly because they see what's working and what isn't. Companies that don't are flying blind.

## WHEN PLANNING MATTERS

I don't want to give the impression that planning is useless. That's not the lesson from submarine hunting. We spent years training before that mission. We had standard tactics. We understood acoustic theory. We knew submarine capabilities. We'd practiced coordination as a crew thousands of times.

**Planning matters for building capability. Action matters for applying that capability to reality.**

The distinction is crucial:

**Plan for:**

- Building skills and capabilities your team needs
- Understanding fundamental principles that won't change
- Establishing coordination mechanisms and communication protocols
- Creating decision-making frameworks
- Preparing for categories of scenarios you might encounter

**Act and adapt for:**

- Applying those capabilities to specific situations
- Dealing with circumstances you didn't anticipate
- Learning which principles matter most in your particular context
- Adjusting tactics based on how the competition/market/enemy responds
- Discovering problems and opportunities that planning couldn't reveal

Google planned extensively for their technical infrastructure—hiring great engineers, building distributed systems expertise, and establishing code review practices. But they adapted constantly on product decisions—launching Gmail as invite-only, adding features based on usage, and killing products that didn't work.

Amazon planned their logistics and fulfillment capabilities—investing in warehouse systems, delivery networks, and inventory management. But they adapted on product selection—starting with books and then expanding based on what customers actually bought and what margins looked like.

**Plan for capabilities. Adapt for application.**

## THE STARTUP THAT WAITED TOO LONG

Let me give you a counter-example: a company that planned when they should have acted, and paid the price.

In 2010, a startup called Color launched with $41 million in funding—one of the largest seed rounds ever at the time. They were building a photo-sharing app with a novel feature: Photos were automatically shared with people nearby based on location. Before launching, they:

- Raised enormous funding to build the "perfect" product
- Hired a large team of engineers and designers
- Spent over a year in stealth mode developing the app
- Designed sophisticated technical infrastructure
- Created extensive marketing campaigns

When they finally launched in March 2011, the app was technically impressive. But users didn't understand it. The value proposition wasn't clear. The sharing model felt weird. Adoption was minimal. Within weeks, it was clear the product didn't work. But because they'd raised so much money and spent so long building, they were committed to the approach. They iterated on the edges—tweaking features and adjusting messaging—but they didn't fundamentally pivot.

By October 2012, Color shut down. They'd burned through $41 million building a product that users didn't want.

**The mistake: They planned and built for over a year without testing their core hypothesis with real users.**

If they'd launched an MVP after three months—just the basic photo-sharing concept, crudely built—they would have discovered immediately that users didn't understand the value proposition. They could have pivoted then, with $38 million still in the bank and time to experiment. Instead, they "planned the perfect hunt" for a year, and when they finally engaged with reality, they discovered their target wasn't where they thought it would be—and they'd used up all their sonobuoys.

## THE PEAR THERAPEUTICS COLLAPSE

Pear Therapeutics stands as perhaps the most expensive lesson in the difference between innovation and commercialization. The Boston-based digital therapeutics company raised over $400 million from prestigious investors including SoftBank Vision

Fund, Temasek Holdings, and Novartis. They went public via SPAC in December 2021 at a stunning $1.6 billion valuation, roughly 400 times their 2021 revenue of approximately $4 million. They were pioneers, becoming the first company to receive FDA approval for a mobile app to treat substance use disorders when reSET was approved in 2017. They had everything that should have mattered: FDA-approved products, clinical evidence of efficacy, and partnerships with major health systems.

Yet just 16 months after going public, Pear filed for Chapter 11 bankruptcy in April 2023. Their entire portfolio of FDA-approved digital therapeutics was auctioned off for a humiliating $6 million, less than 0.4 percent of their peak valuation. The company burned through nearly half a billion dollars without ever finding a sustainable business model. The failure wasn't technological. The technology worked. The clinical evidence was solid. The FDA had validated their approach. The failure was a refusal to adapt to market reality.

The core problem was pricing and reimbursement. Pear's products cost $400 to $500 per patient per month, targeting difficult-to-treat conditions such as substance use disorders and chronic insomnia. When health insurers refused to provide coverage at scale, Pear faced a clear pivot signal. They had several obvious options: Dramatically reduce pricing to drive adoption and prove value at scale, shift to direct-to-consumer models where patients paid out of pocket, or target higher-value use cases where payers would pay premium prices.

Instead, they doubled down on the original strategy. CEO Corey McCann blamed insurers in a LinkedIn post after the bankruptcy filing, stating the company had shown their products could "improve clinical outcomes" and "save payers money," but insurers still denied coverage. This is the founder's trap we've discussed throughout this chapter. When your entire business model depends on someone else changing—in this case, the entire health insurance industry—you've already lost.

The signal was deafening. Despite optimistic projections in late 2022 that 2023 would bring revenues of more than $27 million, sales continued to dramatically underperform. Quarter after quarter, the reimbursement environment didn't improve. Insurers didn't suddenly start covering digital therapeutics at the prices Pear needed. The market was telling them something clear: Your pricing doesn't work, your distribution model doesn't work, and your assumptions about payer behavior are wrong.

Rather than treating this as a mandate to fundamentally rethink the business model, leadership kept raising more money and waiting for payers to "see the light." They believed they were right and the market was wrong. They believed that if they just had more clinical data, more evidence, and more time, the insurers would eventually come around. They burned through hundreds of millions of dollars waiting for a market transformation that never came.

## THE ADAPTIVE ALTERNATIVE

Compare this to Omada Health, a digital therapeutics competitor that achieved profitability by focusing on prevention programs for prediabetes and hypertension—conditions in which employers and payers had clear ROI incentives to invest. Omada's products targeted chronic conditions that were extremely expensive to treat once they progressed. Prediabetes that becomes diabetes costs the healthcare system tens of thousands of dollars per patient. Uncontrolled hypertension leads to heart attacks and strokes. Payers could see the math: Spend a few hundred dollars now to prevent tens of thousands in future costs.

The difference wasn't the technology. Both companies used similar approaches—mobile apps, behavioral interventions, and remote monitoring. The difference was strategic adaptation to market readiness. Omada asked: "Where is the market actually willing to pay today?" not "Where do we wish it would pay?" They found conditions where the ROI was obvious and immediate, where payers didn't need to be convinced or educated, and where the business model could work within the existing reimbursement environment. Pear asked: Where is the clinical need greatest? They focused on substance use disorders and insomnia, devastating conditions with enormous human cost but murky ROI calculations and stigmatized patient populations that insurers were reluctant to invest in. They were clinically right but commercially wrong.

## THE GUILFORD COURTHOUSE LESSON

This is the organizational equivalent of what would have happened if Nathanael Greene had insisted on fighting the British in conventional pitched battles because that's what "real armies" did. He could have spent years waiting for the Continental Congress to give him enough trained regulars, artillery, and supplies to match the British force-for-force. He could have blamed the Continental Congress for not supporting him properly. He could have argued that his strategy was correct and the resource constraints were the problem.

Instead, he accepted the reality he faced. He had militia, not regulars. He had smaller numbers. He had limited supplies. He designed a strategy that worked within those constraints rather than waiting for the constraints to disappear.

Pear Therapeutics refused to accept the reality they faced. They had products that insurers wouldn't reimburse at the prices they needed. They had a distribution model that didn't work. They had a market that wasn't ready for their solution at their price point. Instead of adapting—changing pricing, changing target conditions, or changing their go-to-market strategy—they kept fighting the same battle in the same way, burning through capital while waiting for the market to change.

The market didn't change. The company died. And the technology that might have helped thousands of patients with substance use disorders was sold for pennies on the dollar, a casualty of strategic inflexibility masquerading as principled conviction.

Adaptive intelligence means knowing when you're fighting a battle you can't win and having the courage to retreat to ground where you have advantage. Pear Therapeutics had every opportunity to make that strategic retreat. They chose to stand their ground instead. And they paid the price that every organization pays when it confuses stubbornness with strength.

## INSTAGRAM (REDUX)

Remember Instagram from Chapter 5? The story is even more relevant here.

Burbn—the check-in app that became Instagram—didn't fail because the founders didn't plan enough. They'd built a full-featured app with check-ins, photo sharing, comments, likes, and future plans. It failed because users didn't engage as expected. The planning was fine. The execution was fine. But the market reality was different from their hypothesis. The critical moment wasn't when they decided to pivot to photos-only. The critical moment was when they looked at their usage data and saw that users were engaging with only one feature—photo filters.

**They had data. They looked at it. They adapted immediately.**

If they'd been more committed to their plan—if they'd said, "We need to give check-ins more time to gain traction" or "We just need better marketing"—they would have missed the window. Instead, they killed months of work, focused on the one thing that was working, and rebuilt around it. That decision took weeks, not months, because they were watching real usage data and adapting to what they saw.

**By October 2010—just months after launching Burbn—they'd relaunched as Instagram. Within weeks, they had a million users.** The speed was possible because they were acting and learning, not planning and guessing.

### *The Submarine Commander's Perspective*

There's another dimension to the submarine hunting story worth exploring: the submarine commander's perspective.

The Russian Akula captain we were tracking was doing the same thing we were—making decisions with incomplete information, adapting based on what he observed, and learning in real time. He knew he was being hunted (submarines can detect sonobuoys deploying and often aircraft overhead). So he was adapting:

- Changing depth to exploit thermal layers
- Varying speed to change his acoustic signature
- Going ultra-quiet at times to evade detection

**Both of us were engaged in adaptive intelligence against each other.** This creates an interesting dynamic: In competitive environments, your ability to adapt is relative to your competitor's ability to adapt. It's not enough to be good at learning; you have to learn faster than the competition. In submarine warfare:

- If we drop sonobuoys in a predictable pattern, the submarine commander learns our tactics and evades more effectively next time
- If the submarine uses the same evasion tactics repeatedly, we learn to anticipate and counter them
- Each engagement creates a learning cycle for both sides

**The side that learns faster gains advantage.**

The same dynamic exists in business.**Market Competition Example:**

Company A launches a new feature. Company B copies it three months later. Company A has already learned what works about the feature and has shipped improvements. Company B just copied the initial version, which Company A has already moved past.

Company B looks as if they're following a smart fast-follower strategy. But if Company A is iterating every two weeks based on user data, Company B is always three months behind, copying features that Company A has already improved beyond recognition.

Company A wins not because their initial product was better, but because their learning cycle is faster.

This is why large companies often lose to startups, even when they have more resources: Startups can iterate weekly, while large companies need quarterly planning cycles and approval processes. By the time the large company ships version 1.0 of a feature, the startup has shipped version 4.0.

## ORGANIZATIONAL AGILITY AS COMPETITIVE ADVANTAGE

The ability to run fast learning cycles becomes a sustainable competitive advantage.

Amazon under Jeff Bezos was famous for this. They'd run thousands of A/B tests simultaneously, shipping changes constantly, learning what worked, and immediately implementing it at scale. Traditional retailers couldn't compete, not because Amazon had better initial ideas, but because it learned from millions of customer interactions daily while traditional retailers adjusted their approach quarterly based on sales reports.

The organization that learns faster wins—even if the organization that plans better started with advantages.

### *The Courage to Act*

There's a psychological dimension to "nothing beats action" that's worth examining.

Before we launched from Andøya that night, I didn't know if we'd find the submarine. Our intelligence was good but not certain. Our search plan was reasonable but not guaranteed. Our crew was capable but not infallible. We could have delayed—waited for better intelligence, spent more time planning, or asked for additional assets. But we launched. We engaged. We adapted based on what we found.

**This requires courage: the courage to act despite uncertainty, to commit resources to an approach that might not work, and to make decisions that might be wrong.**

Many leaders lack this courage. They want certainty before they act. They want to know the plan will work before they execute it. They want to eliminate risk before they commit. This is impossible in competitive, uncertain environments. You can't know if the plan will work until you try it. You can't eliminate risk—you can only manage it.

**The leaders who succeed are those who have the courage to act despite uncertainty, coupled with the humility to adapt when their actions reveal new information.**

This isn't reckless action. We didn't just fly randomly around the Norwegian Sea hoping to stumble on a submarine. We had a plan based on the best information available, and we executed it. When reality proved different than our plan, we adapted.

**Courage to act + humility to adapt = adaptive intelligence.**

Many leaders have one without the other:

- Courage without humility = stubbornness (acting but refusing to adapt when proven wrong)
- Humility without courage = paralysis (acknowledging uncertainty but never acting)

You need both.

## THE RED NOSE PHILOSOPHY

That red on our noses represented more than just a successful mission. It represented a philosophy of engagement with reality:

### *1. Training prepares you, but reality teaches you*

We'd trained for submarine hunting for years. But that four-hour mission taught us things training couldn't: how this particular submarine maneuvered, how these particular acoustic conditions affected detection, and how our crew worked together under real pressure. Training builds capability. Reality builds wisdom.

### *2. Plans are hypotheses to be tested*

Our initial search pattern was a hypothesis about where the submarine would be. When that hypothesis proved wrong, we adapted rather than persisting stubbornly. Smart leaders treat their strategies as hypotheses—beliefs to be tested and adjusted, not commandments to be defended.

### *3. Iteration beats optimization*

We didn't spend hours designing the perfect search pattern before our first sonobuoy drop. We dropped a reasonable pattern, evaluated the results, adjusted, and repeated. Each iteration was fast and informed by real data. Together, the iterations were far more effective than a single optimized plan would have been.

### *4. Action creates information*

Until we dropped sonobuoys, we had no information about whether the submarine was in our search area. The act of searching created the data we needed to search more effectively. In business, launching a product creates usage data. Running an ad campaign creates conversion data. Making a hire creates performance data. Action creates information that planning can't generate.

### *5. Success requires teamwork*

No single person found that submarine. The pilots flew the search pattern. The sensor operators analyzed contacts. The in-flight technicians deployed sonobuoys. The navigator maintained the tactical picture. I coordinated the overall effort. Everyone depended on everyone else. Submarine hunting is the ultimate team sport.

Modern product development is the same: designers, engineers, product managers, data analysts, customer support—everyone contributes essential pieces. The teams that coordinate effectively learn faster than those with brilliant individuals working in silos.

## FROM THE NORWEGIAN SEA TO SILICON VALLEY

June 9, 2003. We climbed down from our P-3 with our red noses after thirteen hours in the air—exhausted, exhilarated, and permanently changed. We'd successfully tracked a Russian submarine through multiple evasions and course changes. We'd adapted our tactics in real time based on what we discovered. We'd learned more about submarine hunting in four hours of actual contact than we could have learned in months of simulated exercises.

That red nose on our aircraft was proof: **We'd engaged with reality and won.**

Twenty years later, in boardrooms and startup offices, the same principle applies: **No amount of planning, market research, or strategic analysis can substitute for building something real,**

**putting it in front of real users, and learning from what actually happens.**

The companies that win aren't the ones with the best plans. They're the ones that:

- Act quickly with reasonable hypotheses
- Instrument their products to gather real usage data
- Adapt immediately based on what they learn
- Iterate faster than their competitors
- Have the courage to act despite uncertainty and the humility to change when they're wrong

This is adaptive intelligence: learning by doing, adjusting based on feedback, and iterating toward success rather than planning toward perfection. The 2/47th Infantry learned this in the streets of Cholon when they had to invent urban combat tactics in real time. Nathanael Greene's militia learned this at Guilford Courthouse when they had to execute a retreat strategy they'd never practiced. Combat Air Crew Four learned this in the Norwegian Sea when we had to find a submarine that wasn't where our intelligence said it should be.

And every successful company in the AI era is learning it now: **The only way to adapt to disruption is to engage with it directly, learn from real-world results, and iterate faster than the competition.**

You can plan for months and launch perfectly and fail. Or you can act quickly, learn constantly, and adapt toward success.

**The red-noses in Norway demonstrated which approach works.**

The question is: Are you still planning your hunt, or have you engaged with reality? Because somewhere, your competition has already launched. They're gathering data. They're learning what works. They're iterating. And every day you spend perfecting your plan, they're another iteration ahead of you.

**Nothing beats action.** Go earn your red nose.

# *Section III: Empowered Teams*

# CHAPTER 8

# *The Committee of Safety*

THE COURTHOUSE IN MARTINBOROUGH sat on the ground too new to hold ghosts. Built in 1774 from squared pine logs, it smelled of turpentine and sawdust even in winter, when frost silvered the shakes and made the split-rail fence posts look like old men bent under the weight of coming trouble. On the morning of February 10, 1776, Captain George Moye II pushed through the plank door an hour before the Committee of Safety was set to convene, his boots tracking red clay across the swept floor.

He was fifty-four that year, his beard gone fully gray, his hands rope-scarred from wagon work and tobacco hoeing. The unofficial rank of "captain" had attached itself to him a decade earlier, when Pitt County freeholders chose him to drill Company One of the local militia—not because he'd attended any academy or carried aristocratic blood, but because when a neighbor's barn burned, he showed up with timber and labor; and when disputes arose over property lines, he could read the surveyor's chains and settle matters without lawyers or litigation.

This morning the Committee faced a question that couldn't be settled by common sense alone: David Whitfield's farm had been raided by a band of men calling themselves Loyalists. They'd taken two horses, a wagon, and forty pounds of cured pork. Whitfield wanted the Committee to authorize a pursuit and recovery—armed, if necessary. But the raiders had kin in the county, families who'd made no secret of their conviction that all this talk of Continental Congresses and independence was treason pure and simple. Send militia after them, and you risked neighbor shooting neighbor. Do nothing, and you told every fence-sitter in Pitt that resistance had no teeth.

George Moye understood what the others were only beginning to learn: There would be no royal magistrate riding down from New Bern to adjudicate this. Governor Martin had fled to a British sloop in the harbor. The old order, for all its flaws and rigidities, had at least provided someone to whom hard questions could be deferred. Now the hard questions landed on pine tables in county courthouses, and the men sitting around those tables had to decide with nothing but their judgment and the tentative authority of their neighbors' trust.

By nine o'clock, the other committee members had arrived—planters, a blacksmith, a ferryman, and the Presbyterian minister who doubled as schoolmaster. They settled onto benches, hats in laps, and George called the meeting to order. He laid out Whitfield's complaint in plain terms, then posed the question: "Do we pursue, and if so, under whose authority?"

The discussion that followed would have seemed mundane to a general commanding armies. There was no talk of grand strategy, no dispatches from the Continental Congress. But in that hour of debate—whether to send six armed men after horse thieves who might be neighbors' sons, whether to risk escalating a theft into a blood feud, and whether the Revolution they'd pledged to support could survive without enforcing property rights—the committee was doing the essential work of self-governance. They were learning, in real time, that distributed authority was not permission to act without consequence, but the burden of acting when no higher authority would shoulder the risk.

George Moye let the men talk themselves out, and then he offered his judgment: Send two men to the suspected homestead, unarmed, to request return of the property. If refused, convene the militia not for recovery but for a show of resolve—a drill on

Whitfield's land visible to anyone passing by on the road. Make clear that order would be kept, but don't spill blood unless forced. The committee voted. It carried.

What George Moye had done, without benefit of West Point or parliamentary procedure, was articulate a commander's intent before the term existed: We defend property and order (purpose), we avoid unnecessary violence (constraint), and we trust the men on the ground to execute with judgment (empowerment). The committee adjourned. The recovery party rode out that afternoon. The horses came back three days later, no shots fired. The wagon and pork were forfeit, but the message was sent: Pitt County would govern Pitt County.

## THE ARCHITECTURE OF DISTRIBUTED AUTHORITY

Between 1774 and 1781, Committees of Safety became the shadow government of revolutionary North Carolina. They were not designed by political theorists or modeled on Enlightenment philosophy. They emerged because the collapse of royal authority created a vacuum, and nature—political and otherwise—abhors a vacuum. In Pitt County alone, the committee handled everything a modern county government might: They allocated militia supplies, arbitrated disputes over debts and land, enforced price controls on scarce goods, identified and sometimes detained Loyalist agitators, coordinated with neighboring counties on refugee relief, and conscripted wagons for Continental use.

The genius of the system, if "genius" is not too grand a word for improvisation under duress, lay in its recognition of three principles that would later be codified in military doctrine and, much later still, in the organizational theory of high-performing businesses:

**First, local knowledge beats centralized oversight.** The Committee of Safety in Martinborough knew which families were

Loyalist by conviction and which were merely cautious, which roads flooded in March, which merchants hoarded powder, and which debtors could be squeezed for tax contributions. A directive from the Continental Congress in Philadelphia might call for raising two militia companies; the Pitt committee knew that one of those companies would have to come from the western district where the men spoke a different English dialect and answered to kin networks that predated county lines. Only locals could navigate that terrain.

**Second, shared purpose enables independent action.** The Committees of Safety didn't wait for permission to defend the Revolution. They acted in alignment with a purpose articulated by the Continental Congress and, before that, by pamphlets and sermons and the accumulating weight of grievances: The cause was self-governance, the preservation of property and liberty against arbitrary taxation, and the right of a people to consent to the laws under which they lived. George Moye and his fellow committee members signed the Pitt Association on July 1, 1775—a document that pledged "life and fortune" to whatever course the Congress chose. That pledge was not allegiance to specific orders; it was alignment with a vision. Once aligned, the committee could act on its own authority, knowing its decisions served the larger cause.

**Third, accountability flows from reputation, not rank.** Captain Moye had no commission from a general, no salary, and no legal immunity. If he abused his authority—say, by settling personal scores through militia raids or seizing property for private gain—his neighbors would recall it. Poor judgment destroyed a man's standing faster than any court-martial. This was not accountability in the modern bureaucratic sense, with appeals and procedures and written reprimands. It was older and harsher: A man's word was his currency, and a reputation once spent could not be borrowed back. That knowledge kept the Committees of Safety from becoming petty tyrannies.

In the ledger of historical what-ifs, it's worth considering the counterfactual: What if the Committees of Safety had waited for centralized approval? What if every militia muster, every property dispute, and every wagon requisition required a dispatch to Philadelphia and a written reply? The Revolution would have collapsed under its own bureaucratic weight before the first major battle. The British Empire's great weakness—its insistence on centralized command flowing from London through layers of colonial administration—became the Patriots' great advantage. The Revolution succeeded not despite the absence of top-down control, but because of it.

## JULY 1, 1775: EIGHTY-EIGHT SIGNATURES

Summer in Pitt County meant heat that rose from the fields in visible waves, mosquitoes that bred in standing water, and the constant risk of fever. On July 1, 1775, eighty-eight men gathered in Martinborough under an anvil-bright sun to sign a document that would later be called the Pitt Association. The parchment itself has not survived, but the names were recorded in the colonial secretary's files and the text was later reprinted in the *North Carolina Gazette*. It was not a declaration of independence—that would come a year later—but it was a public commitment, and in the summer of 1775, public commitment carried tangible risk.

Signing meant identifying as a Patriot in a county where Loyalist sentiment ran strong in certain neighborhoods. It meant becoming a target if British regulars or Loyalist militia decided to pacify the backcountry. It meant risking one's tobacco crop if the Royal Navy blockaded the coast and shut down the export trade that turned leaf into coin. It meant a man explaining to his wife why he was wagering the family's future on a cause that had not yet won a single major battle.

Captain George Moye signed first. His son, George Moye III, twenty-three years old that summer, signed second. The act linked two generations in a compact: the elder staking his accumulated property and standing; the younger staking his life expectancy in what promised to be a long and brutal conflict. Eighty-six other men added their names. Some signed with a confident hand, and others with an X-mark witnessed by a literate neighbor. What united them was not ideology in any abstract sense—few had read Locke or Montesquieu—but a shared conviction that they would not be governed without consent, taxed without representation, or subordinated to officials they had no voice in choosing.

The Pitt Association's language straddled two worlds. It pledged loyalty to King George III even as it promised to resist his policies "with our lives and fortunes." This was the hedged bet of men who understood they might lose and wanted language that could be defended as loyal petition rather than open rebellion. But the practical effect was clear: Pitt County would organize its own defense, allocate its own resources, and govern its own affairs. The signatures didn't create that authority—they acknowledged it.

In modern terms, what the Pitt Association represented was a public articulation of purpose. Purpose isn't vision; vision describes a desirable future ("We will be independent"), while purpose explains why the effort matters *now* ("We defend the right of a community to govern itself"). Vision is aspirational and often distant; purpose is immediate and personal. George Moye could rally his militia company not by painting a picture of a post-war republic, but by asking them to defend their neighbors, their farms, and the principle that Pitt County belonged to the people who worked its soil. Purpose converts fear into resolve because it locates meaning in the present struggle, not in a far-off reward.

The Committee of Safety enforced that purpose. When militia drills flagged, George Moye reminded his men that every quarter-hour spent learning to march in formation was an investment in keeping British regulars off their land. When neighbors balked at giving up a wagon for Continental supply runs, the committee invoked the Pitt Association: You signed your name, you pledged your fortune, and a wagon is part of that fortune. Purpose held because everyone had signed. The signatures weren't symbolic; they were social collateral.

## THE MILITIA DRILL: ACCURACY, THRIFT, RESTRAINT

George Moye drilled his militia company twice each quarter, more often than the law required. The muster ground was a cleared acre behind the courthouse, marked at one end by a split-rail fence and at the other by a line of sweetgums. On drill days, men brought their own muskets—Brown Bess surplus if they were lucky, or hunting pieces if they weren't—and lined up in two ranks while George paced the file with a hickory staff, calling commands.

"Shoulder arms. Make ready. Present. Fire!"

But they didn't fire. Powder was scarce and expensive, imported through blockade-runners or hoarded from pre-war stocks. George Moye had a rule, and he repeated it at every drill: "Sight no higher than a squirrel's head on a rail-fence. Pull smooth. We will not waste the King's powder on the King's birds."

The phrase became legend in Pitt County, repeated by later generations as proof that thrift and deadliness were compatible virtues. But what George was teaching went deeper than marksmanship. He was training his men to internalize restraint as a tactical principle. In an era when close-order musket volleys were the norm—mass fire at fifty paces, reload, and fire again—the Patriot militias couldn't match British discipline or firepower. They could, however, conserve ammunition, wait for the decisive shot, and make every round count. At Moore's Creek Bridge and later at Guilford Courthouse, that discipline would prove decisive.

George's other gift to his men was clarity of role. In the militia, every man knew his place in the line, his responsibility in a firing sequence, and the outcome that success required. This wasn't micromanagement; it was structure. Modern organizational theory sometimes conflates empowerment with the absence of structure, as if giving people freedom means removing all constraints. But George Moye's militia proves the opposite: Empowerment requires clear roles and shared understanding of the mission. When you know what's expected, you can act without waiting for orders. When you trust that your neighbor knows his role, you can focus on yours.

By the autumn of 1775, Company One had achieved a reputation. A field return taken on October 12 listed "Capt. George Moy, 58 effectives"—fifty-eight men fit for duty, armed, and trained. That number would fluctuate over the next six years as men rotated between farm and field, but the core remained. They were never the largest company in Pitt County, but they were the most reliable. When orders came to march for Wilmington in February 1776, Company One formed up without complaint. When the Continental Army needed wagons, George's teamsters delivered them. Competence breeds trust, and trust enables delegation.

## MOORE'S CREEK BRIDGE: TESTING DISTRIBUTED COMMAND

The summons arrived at sunset on February 26, 1776: March for Wilmington. Highland Scot Loyalists, armed and organized by Governor Martin, were moving toward the coast to link up with a British expeditionary force. If they succeeded, the Patriots' hold on the North Carolina backcountry would collapse.

What's remarkable about the Patriot response isn't that militia companies marched. It's that they marched without waiting for detailed orders from a central command. The Committee of Safety in Martinborough received a dispatch from the Continental commander requesting reinforcements. Captain George Moye read it to his company, explained the strategic situation, and asked who could march within twelve hours. Fifty-three of fifty-eight men stepped forward. No one needed to tell them what "reinforcements" meant. No one needed to explain why stopping the Loyalists mattered. The purpose was clear, and the decision was theirs.

This is the essential characteristic of empowered teams: They don't need permission to act on shared purpose. They need clarity about the mission and trust that their judgment will be respected.

Company One reached Moore's Creek three days later. When they arrived, Patriot forces under Colonel Richard Caswell and Colonel Alexander Lillington were already entrenched on the far side of the creek. But there was no unified command structure in the modern sense. Caswell commanded the Dobbs County militia. Lillington commanded the New Hanover County men. George Moye commanded Pitt County's Company One. They weren't integrated into a single regiment with a clear chain of command. They were independent units aligned around a common purpose: Stop the Highlanders from reaching the coast.

The night before the battle, the commanders met—not to issue orders, but to coordinate. Caswell proposed greasing the bridge planks. Lillington suggested removing planks entirely to create gaps. George Moye, the most experienced in defensive tactics, recommended both: Remove enough planks to slow the charge, and grease what remained to make footing treacherous. Each commander then returned to his own unit to prepare.

There was no written battle plan, and no detailed orders about fields of fire or fallback positions. What there was instead was the commander's intent: We hold the far side of the creek, we funnel the Highlanders onto the bridge, and we wait until they're committed before opening fire. Each company commander understood the intent and could make tactical decisions within that framework.

When the Highlanders charged at dawn, Company One was positioned in reserve on the right flank. George Moye III, serving as a sergeant, watched the Highland column advance onto the greased planks, watched them slip and stagger, and watched the Patriot artillery open fire from entrenched positions. The battle was over in fifteen minutes, won by militia companies acting independently within a shared strategic framework.

## WHAT MADE IT WORK

Company One never fired a shot. They spent the morning ferrying prisoners and inventorying captured weapons. In a traditional military structure, this would have felt like missing the glory. But George Moye understood something his men had learned through the Committee of Safety experience: every role matters when everyone understands the mission.

The militia privates guarding prisoners weren't just following orders. They understood that securing eight hundred captured Loyalists meant eight hundred fewer fighters available to British forces. The teamsters logging muskets and powder horns understood that those weapons would equip under-supplied militia companies across North Carolina. Purpose converted logistics into meaningful contribution.

This is what distinguishes empowered teams from merely delegated teams. Delegation says, "Do this task." Empowerment says, "Here's why this matters—now execute with judgment."

After the battle, Colonel Caswell wrote a dispatch to the Continental Congress describing the victory. He didn't mention unified command or brilliant generalship. He wrote: "The several militia companies acted with uncommon spirit and in good order, each commander exercising judgment suited to his position." In modern terms, he was describing a distributed authority structure that succeeded precisely because it was distributed.

## GUILFORD COURTHOUSE: PRE-DECLARED RETREAT AND TACTICAL AUTONOMY

Five years later, the principles tested at Moore's Creek would face their ultimate examination at Guilford Courthouse. By March 1781, General Nathanael Greene commanded a Continental force reinforced by militia companies from across the Carolinas. He faced Cornwallis's professional British army. They were better trained, better equipped, and accustomed to winning.

Greene's battle plan was unorthodox: He would deploy in three lines, with the militia in the front two lines instructed to fire volleys and retreat. This wasn't a contingency plan for if things went badly. This *was* the plan. The militia would inflict casualties and then withdraw in order, preserving Greene's force while bleeding the British.

For this to work, every militia officer down to the platoon level had to understand not just what to do, but why. Greene didn't micromanage the execution. He communicated intent and trusted his subordinate commanders to implement it.

Lieutenant George Moye III commanded a platoon in the second line. The night before the battle, his father—now Captain George Moye, age fifty-nine and suffering from gout—walked the line, explaining the strategy to junior officers and sergeants. He was not issuing detailed instructions, but ensuring they understood the logic: "We are not holding ground. We are breaking their strength. Two volleys, three if time allows, and then fall back in order. Cornwallis can take the field. He cannot afford to lose a quarter of his army doing it."

This kind of commander's intent briefing is standard doctrine in modern military operations, but in 1781 it was revolutionary. The British model assumed officers would execute orders without needing to understand the broader strategy. Greene's model assumed that understanding the strategy was what enabled disciplined execution.

## PURPOSE UNDER FIRE

When the British columns advanced at dawn on March 15, the North Carolina militia in the first line fired their two volleys and retreated exactly as planned. No general stood behind them, threatening to shoot deserters. No officers screamed orders to hold ground. They executed the retreat because they understood it was the plan, not a failure.

In the second line, George Moye III's platoon faced the British grenadiers closing to fifty paces. Musket balls snapped branches overhead. A soldier to his left took a ball through the shoulder and crumpled. Another started to break ranks.

George grabbed the man's collar and said, "Purpose, lad, purpose!"

It's a brief moment in the historical record, but it captures everything about empowered teams. George wasn't telling the private what to do. The private knew what to do. He was reminding him why. The platoon wasn't holding ground because a general somewhere had decided this patch of North Carolina dirt mattered. They were holding long enough to inflict maximum casualties and then falling back to preserve the army. Every private in that line understood the mission.

The second line fired their third volley at thirty paces, close enough that the British line staggered. Then they retreated in formation, falling back to Greene's third line. The battle raged for another two hours. Cornwallis won the field but lost over five hundred men, a quarter of his force. Greene retreated with his army intact. Within months, Cornwallis would be trapped at Yorktown, his force too depleted to hold the Carolinas.

## THE EMPOWERMENT ARCHITECTURE THAT MADE IT POSSIBLE

Guilford Courthouse succeeded as a strategy because Greene had built an empowerment architecture into his command structure:

**Clear commander's intent.** Every officer understood the strategic objective: Inflict casualties, preserve the army, and make the British victory pyrrhic. This clarity allowed tactical flexibility. When George Moye III's platoon faced heavier pressure than expected, he could decide whether to fire the third volley or pull back immediately, because he understood the intent was to bleed the British, not to die holding ground.

**Trust in subordinate judgment.** Greene didn't position himself in the second line micromanaging platoon movements. He communicated the plan and trusted his junior officers to execute it. When circumstances changed—when the British advanced faster or slower than expected, and when militia units on the flanks wavered—those officers had the authority to adapt without requesting permission.

**Accountability through reputation.** The militia officers knew their decisions would be scrutinized afterward, not by generals reading reports, but by the men they commanded and the communities they returned to. This wasn't accountability enforced through courts-martial—it was accountability enforced through reputation. If George Moye III had panicked and broken early, he would have answered for it to his platoon and his neighbors for the rest of his life. That knowledge created a different kind of discipline than fear of punishment.

**Shared sacrifice creating cohesion.** The men in George's platoon weren't strangers drafted into service. They were neighbors who'd signed the Pitt Association together, drilled together on the courthouse grounds, and marched to Moore's Creek together. They'd shared purpose for six years. That history created trust that no amount of training could manufacture quickly.

Compare this to the British structure. Cornwallis commanded professional soldiers who executed orders brilliantly. But those soldiers didn't understand why they were fighting in North Carolina, didn't know the strategic objective beyond "take the field," and couldn't adapt when the Patriots refused to fight the kind of battle the British were trained for. The British won the battle tactically. They lost the war strategically, in part because their centralized command structure couldn't adapt to an enemy that operated through distributed authority.

## THE POST-WAR INSTITUTIONAL LEGACY

When the war ended, the Committee of Safety dissolved. But the organizational principles it had established persisted. George Moye received a land grant and divided it among his children and the formerly enslaved people he'd manumitted. This wasn't sentimentality. It was applying the same principle that had governed the Committee of Safety: Stakeholders govern better than subjects.

George Moye III served as Pitt County sheriff, mediating disputes that once would have gone to royal magistrates. He didn't rule by decree. He convened interested parties, established a shared purpose (both sides want a fair resolution), and trusted people to negotiate solutions. The skills learned in militia command—communicating intent, empowering others to act, and holding people accountable through reputation—transferred directly to civilian governance.

The empowered team structure that won battles built institutions that lasted—not because the structure was imposed from above, but because the people who'd learned to govern themselves through Committees of Safety carried those principles into peacetime. They'd proven that distributed authority aligned around clear purpose could defeat centralized command. And they built a republic on that proof.

## THE MODERN PARALLEL: NVIDIA AND THE FLAT HIERARCHY

In 2024, NVIDIA became one of the world's most valuable companies, with its market capitalization exceeding $3 trillion at its peak. The company's dominance in artificial intelligence computing—its GPUs power the vast majority of AI training and inference workloads—stems partly from technical brilliance, but also from an organizational structure so unusual it violates nearly every principle of modern corporate management.

Jensen Huang, NVIDIA's CEO and co-founder, has sixty direct reports. This is not a typo, nor is it a temporary arrangement during a transition. Sixty executives report directly to Huang with no intervening layers of senior vice presidents, group presidents, or other buffering roles. In a typical Fortune 500 company, a CEO might have eight to twelve direct reports, which is more than that is considered unmanageable.

Huang explains the logic in interviews: "I want to be in direct contact with the people doing the work. I don't want information filtered through layers. If there's a problem, I want to know immediately, not after it's been sanitized through three management reviews."

The structure works because Huang has created what he calls a "loosely coupled, tightly aligned" organization. "Loosely coupled" means teams have autonomy over execution; they don't wait for permission to move forward, don't require sign-offs on tactical decisions, and don't escalate routine problems. "Tightly aligned" means everyone understands the company's strategic purpose: Dominate AI computing by delivering performance that competitors can't match.

Huang achieves alignment not through org charts but through relentless communication. He writes detailed technical emails to the entire company explaining strategic decisions. He holds no one-on-one meetings with executives—all communication happens in group settings where everyone hears the same message. He encourages anyone in the company to email him directly with ideas or concerns. The goal is shared context: If all employees understand *why* NVIDIA is pursuing a particular architecture or market, they can make tactical decisions that align without needing approval.

What Huang has built, whether he'd describe it this way or not, is a modern instantiation of the Committee of Safety model. NVIDIA's teams function like Pitt County's militia companies: They have clear purpose (dominate AI computing), local authority (teams own their technical domains), and accountability (results are visible, performance is measured, and reputation matters). The company moves faster than competitors because decisions are made by the people closest to the work, not by executives insulated from technical reality.

Consider the contrast with a traditional hierarchical model:

| Centralized Hierarchy | NVIDIA's Flat Model |
|---|---|
| Decisions require approval up the chain | Teams empowered to execute within strategic bounds |
| Information filtered through layers | Direct communication from CEO to engineers |
| Title-based authority | Competence-based influence |
| Slow adaptation to market shifts | Rapid pivots based on ground-level insight |
| One-on-one meetings create information silos | Group meetings ensure shared context |
| Risk-averse (approval culture) | Risk-tolerant (trust-based culture) |

The parallel to revolutionary Committees of Safety is precise: Both systems empower local actors to make decisions in alignment with shared purpose. George Moye's militia company didn't wait for Continental Congress approval to drill or to requisition supplies; they acted within understood constraints. NVIDIA's engineering teams don't wait for VP approval to pursue technical improvements; they act within strategic direction.

The danger of centralized control, whether in colonial administration or corporate bureaucracy, is paralysis. When every decision requires approval, decision-making becomes a bottleneck. The organization moves at the speed of its slowest approval process. By the time a decision reaches the top, the market has shifted, the technical landscape has changed, or the opportunity has closed.

Huang's sixty direct reports would be chaos if NVIDIA lacked clarity of purpose. But because the company's mission is articulated relentlessly—in emails, in all-hands meetings, and in technical reviews—executives can act independently while remaining aligned. This is commander's intent before it became doctrine: **Here's where we're going and here's why it matters; now execute with judgment.**

The comparison isn't perfect. NVIDIA operates in a competitive market, not a revolutionary war; its failures result in lost revenue, not lost lives. But the leadership principle is identical: Clarity of purpose, distributed authority, and trust in competence beat centralized control when speed and adaptation matter.

## THE LESSONS: EMPOWERMENT BY NECESSITY

The Pitt County militia and the Royal Army represented two different philosophies of organization. The British model was hierarchical, disciplined, and centralized. Orders flowed from generals through colonels through captains through sergeants to privates. Deviation from orders was punishable. Initiative was discouraged. The system worked brilliantly when communications were intact and the chain of command functioned. But the American theater—vast distances, poor roads, and unreliable couriers—disrupted that chain constantly.

The Patriot militia model emerged not from theory but from necessity. There was no standing army, no reliable supply chain, and no professional officer corps. Committees of Safety governed because no one else would. Militia captains led because communities trusted them. Orders were guidelines, not scripts, because by the time a messenger brought word from headquarters, the situation on the ground had changed.

What the Revolution proved, and what modern organizations are still learning, is that empowerment isn't a luxury for stable times; it's a necessity for turbulent ones. When the pace of change exceeds the speed of centralized decision-making, the only viable model is distributed authority aligned by shared purpose.

George Moye II couldn't micromanage fifty-eight militiamen scattered across Pitt County. He couldn't be present at every drill, every skirmish, or every committee meeting. What he could do was communicate purpose—why they were fighting, what success looked like, and what principles they couldn't violate. Once that purpose was clear, his men could act independently. When George III took a musket ball at Guilford Courthouse and still held his platoon in line, he wasn't following his father's orders; he was embodying his father's intent.

The Committees of Safety won the Revolution not because they were better warriors than the British regulars—man for man, they weren't—but because they could adapt faster. They made decisions at the speed of local knowledge. They allocated resources based on immediate need rather than distant policy. They held leaders accountable through reputation rather than rank.

In modern terms, they "out-learned" their opponents. Jensen Huang's phrase—that NVIDIA seeks to "out-learn competitors, not out-decide our organization"—would have made sense to George Moye. You can't out-decide a market that moves faster than your approval process. You can out-learn it only by empowering the people closest to the customer, the technology, and the tactical reality to act on what they know.

The British lost an empire because they couldn't adapt their command structure to the American terrain. Companies lose market share for the same reason: They can't adapt their approval processes to the speed of technological change. The principle is timeless. Clarity of purpose enables speed of action. Distributed authority beats centralized control when the environment demands adaptation.

## EPILOGUE: THE RIVER AND THE COVENANT

The Tar River still runs black through eastern North Carolina, slower now behind dams and flood-control levees, but unmistakable in its dark tannin sheen. The courthouse in Martinborough is long gone, replaced by brick and then by modern concrete, but the crossroads remain. If you drive through Pitt County today, past the tobacco fields now subdivided for housing and the family farms that are now corporate operations, you won't see any monument to the Committees of Safety. There's no bronze plaque commemorating George Moye II's militia drills or the eighty-eight signatures on the Pitt Association.

But the lesson endures in a thousand invisible ways:

- In the town councils that still meet to debate local issues without waiting for state approval
- In the volunteer fire departments that organize themselves and fund their equipment through community support
- In the small businesses that succeed because the owner trusts employees to serve customers without checking every decision
- In the high-performing teams at technology companies where engineers ship features without VP sign-off because everyone understands the product visionThe lesson is this: **Empowerment is not the absence of leadership; it's the highest expression of it.** George Moye led by providing clarity—clarity of purpose, clarity of constraints, and clarity of what success looked like. Once that clarity existed, his men didn't need him to tell them how to fight. They needed him to tell them *why* to fight, and then to trust them to figure out the *how*.

The Moyes, father and son, survived the Revolution because they understood something the British command never grasped: Control is brittle; trust is resilient. Hierarchies break when the leader falls or the communication line severs. Purpose endures because it lives in the minds of everyone who shares it.

The river keeps its silent covenant, bearing the reflected stars of a republic the Moyes helped imagine. They built it not with grand strategy, but with a hundred small decisions made in courthouses and militia grounds and on muddy fields where frightened men looked to their neighbors and asked, "What do we do now?" And the answer, time and again, was: "We know our purpose. Now we act."

That answer echoes still, from Pitt County to NVIDIA's headquarters, wherever leaders trust their people to do what needs to be done without waiting for permission that may never come.

# CHAPTER 9

# *When the Captain Falls*

## THE MATHEMATICS OF SURVIVAL

ON MAY 16, 1863, at Champion Hill, Mississippi, the 57th Regiment Georgia Infantry suffered catastrophic losses: 40 killed, 96 wounded, and 48 captured—a 44 percent casualty rate in a single day's fighting. Private John E. Moye, thirty-eight years old, survived without injury.

The regimental records tell us what happened but not how. We know Colonel William Barkuloo commanded the regiment that day and survived, only to be captured at Vicksburg seven weeks later. We know the 57th Georgia, part of Brigadier General Alfred Cumming's brigade, held a ridgeline against Union General Ulysses S. Grant's forces. We know they fought hard—General Cumming later praised the regiment's actions—and we know they ultimately broke under pressure and fell back toward Vicksburg. What we don't know, and what the sparse records cannot tell us, is exactly how ordinary soldiers, including John E. Moye, functioned when the command structure fractured under fire.

But we can infer from what came next.

Civil War infantry regiments operated with officer casualty rates that would be unthinkable in modern militaries. Officers stood at the front of their companies, marked by swords and sashes, making conspicuous targets. The mathematics were brutal: In major engagements, regiments losing 40 percent of enlisted men typically lost 60–70 percent of their officers. When a regiment suffered 44 percent casualties in a single day, as the 57th

Georgia did at Champion Hill, the leadership structure didn't just thin—it collapsed.

What happens to an organization when the people who give orders are suddenly gone?

The answer depends on what the remaining members understand about their mission. If only captains know the objective, and the captains fall, the unit disintegrates. If corporals and privates also understand what they're trying to accomplish, the unit adapts. Someone steps forward—not necessarily through formal promotion, but through the simple recognition that someone must—and the mission continues.

This isn't heroism. It's organizational necessity.

John E. Moye's survival of Champion Hill without promotion or commendation suggests he wasn't exceptional. He was, by all available evidence, ordinary—a farmer from Washington County doing what tens of thousands of other Confederate soldiers did. But his regiment's continued functioning after losing nearly half its strength suggests that somewhere in the ranks, ordinary men learned to exercise extraordinary responsibility when circumstances demanded it.

## THE FORTY-SEVEN DAYS

After Champion Hill, what remained of the 57th Georgia—342 men out of the original 400—retreated to Vicksburg and prepared for siege. Union General Grant, having failed to take the fortress city by storm, surrounded it with 70,000 men and waited. Inside, 30,000 Confederate soldiers and 3,000 civilians began a 47-day descent into hunger, disease, and despair.

Regimental records from the Vicksburg siege are sparse. Siege conditions collapsed administrative functions—who files paperwork when artillery is destroying your headquarters daily? What we have are fragments: official reports noting that the 57th

Georgia manned the eastern perimeter, casualty lists, and occasional mentions in higher command documents.

One such mention appears in Major General Carter L. Stevenson's after-action report. He noted that Lieutenant-Colonel Cincinnatus S. Guyton of the 57th Georgia led a night sally with portions of his regiment, charging entrenched Union positions on three separate points near Hall's Ferry Road. The attack succeeded, taking the Union trenches "with admirable gallantry" and inflicting "considerable loss to the enemy."

For such an operation to work during a siege that had killed or incapacitated most officers, junior leaders must have taken initiative far beyond their formal rank. Night assaults require coordination, timing, discipline under fire, and the ability to execute complex maneuvers in darkness without constant supervision. Lieutenant-Colonel Guyton could not have been everywhere simultaneously. He needed junior officers, sergeants, and even privates who understood the objective well enough to adapt tactics on the ground.

Whether John E. Moye participated in that particular sally, we cannot know. His name appears nowhere in the reports. What we know is that he survived the siege and was listed on the July 4, 1863, parole rolls as a private when Vicksburg surrendered. That he survived doesn't tell us whether he led or followed, or whether he showed initiative or simply obeyed orders.

But organizational theory tells us something the individual record cannot: For a unit to maintain cohesion through a 47-day siege with dwindling rations, constant bombardment, epidemic disease, and decimated leadership, informal authority structures must fill the gaps left by casualties. Someone has to organize picket rotations. Someone has to distribute rations equitably. Someone has to maintain morale when the situation is obviously hopeless. Someone has to settle disputes before they become desertions.

These functions don't require rank. They require competence and the trust of peers.

The Confederate Army's administrative apparatus had largely collapsed by the summer of 1863. Official promotions took months to process through Richmond. Regiments couldn't wait for paperwork when a sergeant died and someone needed to command his squad that afternoon. So they improvised—conferring authority based on who could exercise it rather than who held the proper commission.

This is distributed leadership by necessity rather than design. And it's what kept the 57th Georgia and other such units functional when formal command structures failed.

## ANDERSONVILLE: THE MORAL LABYRINTH

After Vicksburg fell, Confederate prisoners were paroled—released on their word not to take up arms until officially exchanged. John E. Moye returned to his family in Washington County, Georgia, planted what crops he could manage with most of his field hands gone, and waited for the exchange order. It came in late 1863. The 57th Georgia was reassembled, understrength and exhausted, and sent not to the front lines but to a post that would haunt the men who served there: guard duty at Andersonville Prison.

Andersonville, officially Camp Sumter, was designed to hold 10,000 Union prisoners of war. By spring 1864, it held 33,000 in twenty-six acres. There were no barracks, no latrines, and no clean water supply. Men lived in makeshift tents or dug burrows. The stream running through the camp served as both sewer and water source. Dysentery, scurvy, and gangrene killed 13,000 men—nearly a third of all who passed through the camp—over fourteen months.

The 57th Georgia's role was perimeter security: manning stockade walls, patrolling the "deadline" fence beyond which prisoners would be shot, and guarding the main gate. They did not administer the camp; that fell to Confederate Army quartermasters and provost marshals. But they witnessed the conditions, and they were complicit by presence if not by intent.

Records show John E. Moye served at Andersonville for three months in 1864. There are no surviving letters or diaries describing his thoughts on the assignment. We don't know if he felt moral conflict about guarding men dying of neglect. We don't know if he tried to help prisoners or simply followed orders. What we know is that he completed his service there and later rejoined the 57th Georgia for the Atlanta Campaign—he neither deserted nor refused duty, which suggests he accommodated himself to the situation somehow.

Andersonville appears in this chapter not because John was responsible for the atrocities there—he wasn't, not individually—but because it illustrates an uncomfortable truth about distributed authority: It can serve immoral ends as efficiently as moral ones.

The same decentralized structure that allowed militia companies to defend Pitt County in 1776 also allowed Confederate prison guards to enforce a system that killed thousands through neglect. Empowerment is a tool. Like any tool, its value depends on the purpose to which it's applied.

When junior leaders understand their mission as "keep prisoners from escaping" but receive no guidance on humane treatment, no constraints on acceptable methods, and no accountability for outcomes beyond preventing breakouts, the result is Andersonville. Purpose must be paired with values. A Commander's intent must specify not only what to achieve but also what lines cannot be crossed.

The Union command faced similar moral failures at Northern prisons such as Elmira, where Confederate prisoners died at nearly comparable rates. In both cases, guards were empowered to maintain security but were given insufficient resources and no ethical framework beyond "Don't let them escape."

This is the dark side of distributed leadership. When you push decision-making authority down to junior leaders without also pushing down clear ethical constraints and adequate resources, you create conditions where well-intentioned individuals can participate in systems that produce atrocity.

John E. Moye left Andersonville in summer 1864 when the 57th Georgia was recalled to the front. Sherman's army was driving toward Atlanta, and every available regiment was needed. Whether John felt relief at leaving or guilt at what he'd witnessed, we don't know. What we know is that he marched north to face Sherman, where the organizational skills required for distributed leadership would be tested again.

## KENNESAW MOUNTAIN AND ATLANTA: ORGANIZATIONAL RESILIENCE UNDER FIRE

By June 1864, Union General William T. Sherman had pushed Confederate forces under General Joseph E. Johnston back to a line of hills north of Atlanta. One of those hills, Kennesaw Mountain, offered commanding views of the approaches to the city. Johnston fortified it with artillery and entrenchments.

On June 27, Sherman ordered a frontal assault. It was a disastrous decision—Union soldiers advancing uphill into prepared defenses suffered 3,000 casualties in a few hours while inflicting only 800 on defenders. But for the 57th Georgia, positioned on Pigeon Hill just south of the main Kennesaw peak, the battle validated everything they'd learned through three years of hard fighting.

Official reports note that the regiment held its section of trench line and repulsed multiple Union assaults with minimal casualties—only about a dozen men lost, mostly to artillery fire. What the reports don't capture is how a regiment that had lost its original officer corps at Champion Hill, survived a siege at Vicksburg, and endured months at Andersonville could still function as a cohesive unit.

The answer lies in what organizational theorists call "institutional memory distributed across personnel." The 57th Georgia in 1864 was not the same unit that left Georgia in 1862. Most of the original enlisted men were dead, wounded, captured, or deserted. Yet the regiment still knew how to fight effectively—knowledge had transferred from veterans to replacements, from sergeants to privates, and from experienced soldiers to new conscripts.

This kind of knowledge transfer doesn't happen automatically. It requires veterans who can articulate what they've learned, not just demonstrate it. It requires junior leaders who understand not just their immediate tasks but also the broader context—why they're doing what they're doing, how it fits into the overall defensive scheme, and what alternatives exist when circumstances change.

By Kennesaw Mountain, the 57th Georgia had developed what modern organizations would call "bench strength"—the depth of capability below the starting lineup. When officers fell, junior leaders could step up because they'd been watching officers make decisions for three years. They'd internalized not just what to do but how to think about what to do.

The regiment fought through the rest of the Atlanta Campaign: the Battle of Atlanta on July 22, where Confederate General John Bell Hood's counterattack failed; the Battle of Jonesboro in late August, where Union forces cut the railroads supplying Atlanta; and the evacuation of the city on September 2. Through it all, casualties continued to mount. By the time Atlanta fell, the regiment had cycled through multiple commanders.

John E. Moye, now forty years old and worn down by three years of war, remained with the regiment. Records list him consistently as a private, never promoted. This suggests he wasn't a standout soldier. But it also suggests he was reliable enough to keep around—armies under pressure don't retain dead weight.

His survival through campaigns that killed so many others suggests competence at the fundamental tasks of soldiering: following orders, keeping equipment functional, not getting separated from the unit, and possibly—though we cannot know—exercising judgment in moments when orders were absent or unclear.

## THE LONG RETREAT AND SURRENDER

The 57th Georgia spent the winter of 1864–1865 retreating through Georgia and into the Carolinas, merging with other shattered units to form composite regiments that existed more on paper than in reality. By March 1865, at the Battle of Bentonville, North Carolina, the 57th could muster fewer than a hundred men. They fought one last engagement against Sherman's overwhelming forces, held for a day, and retreated again.

On April 26, 1865, General Joseph E. Johnston surrendered the remnants of the Army of Tennessee at Greensboro, North Carolina. John E. Moye, listed on the parole rolls as a sergeant in the 1st Georgia Consolidated Infantry (the administrative merger of several decimated regiments), stacked his rifle and walked home.

The fact that he's listed as a sergeant on the surrender rolls—after appearing as a private throughout most of the war—suggests that by the end, someone recognized his competence with a field promotion. Or perhaps it was simply record-keeping chaos, and the promotion was never official. We can't know. What we do know is that he enlisted in October 1861 and surrendered in April 1865, serving through Champion Hill, Vicksburg, Andersonville, Kennesaw Mountain, Atlanta, and the retreat through the Carolinas.

He'd seen multiple regimental commanders killed or wounded, served under dozens of different officers, and learned that when the man giving orders falls, the mission doesn't stop—it just transfers to whoever has the competence to pick it up.

## THE MODERN PARALLEL: THE GREAT TECH LAYOFFS OF 2022–2023

In late 2022, the technology industry entered a period of mass retrenchment. Meta laid off 11,000 employees in November and then another 10,000 in March 2023. Amazon cut 27,000 jobs across two rounds. Google eliminated 12,000 positions. Twitter, under new ownership, fired more than half its workforce in weeks. Stripe, Salesforce, Microsoft, Coinbase, DoorDash—the list of companies conducting mass layoffs read like a roster of tech's elite.

The causes were varied: pandemic hiring had outpaced sustainable growth, rising interest rates made growth-at-all-costs strategies untenable, and revenue projections proved too optimistic. But the effect was uniform: Companies suddenly found themselves operating with 10, 20, or 30 percent fewer employees than planned. Critical projects lost key engineers. Teams were disbanded mid-sprint. Managers found themselves supervising twice as many reports. Institutional knowledge walked out the door.

What happened next separated companies that survived from those that floundered. The difference wasn't financial reserves or market position—though those helped—but the depth of empowered leadership below the executive tier.

Companies with strong middle management, with engineers and product managers who understood not just their tasks but also the broader mission, adapted quickly. Companies that had concentrated decision-making at the top, where every feature needed

VP approval and every budget line required executive sign-off, found themselves paralyzed.

Consider two companies, both in the cloud infrastructure space, both conducting roughly 15 percent layoffs in early 2023. These are composites based on multiple real cases, not specific firms, but the patterns are documented.

**Company A** had built a culture of centralized control. Product roadmaps were set by executives; engineers implemented specs. When layoffs hit, Company A lost several senior product managers who'd been the sole keepers of strategic context. Mid-level engineers didn't know why they were building certain features or how those features fit into the business model. When customer priorities shifted mid-quarter—as they often do—there was no one with authority to reprioritize. Work continued, but it was inertial work, finishing projects that no longer served business needs because no one below director level felt empowered to say, "This doesn't make sense anymore."

**Company B** had invested years in what they called "context distribution"—making sure everyone, down to individual contributors, understood the company's business model, revenue sources, competitive threats, and strategic priorities. When Company B conducted layoffs, they lost headcount but not capability. Mid-level engineers could make roadmap calls because they understood which features drove revenue. Product managers could negotiate directly with customers because they knew the cost structure and could make pricing decisions on the spot. Directors could reprioritize entire quarters without executive review because they'd internalized the company's strategic intent.

Company A's product velocity dropped 40 percent in the six months following layoffs. They missed commitments, lost customers, and entered 2024 in turnaround mode. Company B's velocity dropped 15 percent—proportional to headcount lost—but recovered within a quarter. They maintained customer commitments, shipped new features, and emerged from layoffs with market share gains.

The difference was empowerment depth. Company A had empowered executives but not middle managers, middle managers but not senior engineers. Company B had pushed empowerment down to every level by pushing context down to every level. When leaders left, the people remaining knew enough to step up.

My own experience as CEO at Paige illustrated this principle. Paige was a health tech startup spun out of Memorial Sloan Kettering Cancer Center, applying artificial intelligence to cancer pathology. Like many startups, we overhired during the early funding rounds when capital was cheap. When interest rates spiked and the venture capital winter set in, we had to conduct multiple rounds of layoffs over three years—ultimately reducing headcount by over 70 percent.

I hated every layoff. These were talented people, many of whom had taken risks to join a startup, and we were letting them go for reasons unrelated to their performance. But the layoffs were necessary for the company's survival.

What surprised me was that after the layoffs, the remaining teams actually increased productivity. The lack of organizational layers meant the people doing the work also understood the mission directly. Communication loops shortened. Decisions happened faster. We'd inadvertently created the conditions for distributed leadership: small teams with high context and clear objectives.

The parallel to the 57th Georgia isn't perfect—no one dies in a tech layoff, despite what LinkedIn posts might suggest—but the organizational principle is identical. When crisis removes leaders suddenly, the only organizations that survive are those where leadership capability exists at multiple levels. If only captains know the mission, and the captains fall, the regiment breaks. If sergeants and corporals also understand the mission, the regiment adapts.

## SUCCESSION PLANNING AS DISTRIBUTED LEADERSHIP

The modern corporate obsession with succession planning typically focuses on the C-suite: Who replaces the CEO? Who takes over if the CFO leaves? These are important questions, but they miss the deeper challenge. The real succession question isn't "Who's next?" but "Can we function if multiple leaders leave simultaneously?"

The 57th Georgia at Champion Hill lost officers and senior NCOs across multiple companies in the span of three hours. If leadership capability had been concentrated in those men alone, the regiment would have disintegrated. The reason it didn't was that junior leaders—corporals, senior privates, and men who'd been in the ranks long enough to understand what they were trying to accomplish—could step forward and keep the unit functioning.

This kind of distributed leadership requires three things:

### *1. Clarity of Mission Down to the Lowest Level*

The Confederate regiments that held together longest were those whose soldiers understood not just "follow orders" but also "hold this position until ordered to fall back" or "maintain contact with the enemy while the wagon trains withdraw." The mission was clear enough that when officers fell, someone else could execute.

In modern organizations, this translates to making sure individual contributors understand not just their tasks but also the purpose behind them. If an engineer knows the feature they're building is meant to reduce customer churn by improving onboarding, they can make intelligent design decisions even if the product manager is gone. If they know only to "build the feature," they're lost when the PM leaves.

### *2. Training Judgment, Not Compliance*

John E. Moye's ability to function effectively (assuming he did—remember, we're inferring from organizational survival) wasn't because he'd memorized regulations. It was because he'd watched officers make decisions for three years and internalized the judgment frameworks they used. How do you allocate scarce resources fairly? How do you rotate men through dangerous duty? How do you maintain morale when the situation is hopeless?

Modern organizations train compliance—here's the policy; follow it—but rarely train judgment. Judgment requires exposure to real decisions, explained reasoning, and permission to make small mistakes in low-stakes situations so you don't make large mistakes when it matters.

### *3. Acceptance of Temporary Hierarchy*

When Confederate regiments needed someone to function as a sergeant, they didn't wait for promotion orders from Richmond. They said, "You're acting as sergeant," and the person acted accordingly. The Confederate Army's administrative structure was flexible enough—or broken enough—to allow informal rank based on competence.

Modern organizations, especially large ones, often lack this flexibility. A senior engineer can't suddenly start making product decisions even if he's the only one left who understands the codebase; that's not his role. A middle manager can't reprioritize a division's roadmap even if all the VPs are gone; that exceeds his authority.

This rigidity kills adaptability. Organizations that survive crisis give temporary authority to whoever can exercise it competently, and then ratify or adjust later.

## THE MORAL LEDGER

Before moving to business implications, we must address the uncomfortable fact that John E. Moye fought for the Confederacy, an entity whose core purpose was preserving slavery. Records suggest his family owned enslaved people. He served at Andersonville, where thousands died from neglect.

How do we extract leadership lessons from a man serving an immoral cause?

The answer lies in separating personal leadership capability from organizational purpose. John E. Moye's ability to function when formal authority structures collapsed, to maintain unit cohesion under stress, and to operate with incomplete information—those were his skills. Skills can serve good ends or bad ends. The same qualities that made him effective in a Confederate regiment would have made him effective in a Union regiment or in any organization facing crisis.

Leadership literature often sanitizes history, selecting only examples from causes we now judge righteous. But leadership skill is morally neutral. Rommel was a gifted battlefield commander; that doesn't make Nazism defensible, but it also doesn't mean we can't study Rommel's tactics. John E. Moye's service in a slave-owning army doesn't mean his adaptability is worthless as a case study; it means we must be clear about context.

The lesson from the 57th Georgia isn't "Fight for lost causes with competence." It's "Organizations facing existential crisis survive when ordinary people can exercise leadership competence." That principle applies whether the cause is defending slavery (indefensible) or defending South Korea from invasion (defensible) or surviving corporate bankruptcy (morally neutral). The skill set is separate from the cause.

This distinction matters because leadership is, ultimately, a human capability that exists in every society, serving every kind of purpose. If we study only those leaders whose causes we approve of, we miss the full range of how leadership functions under stress.

## THE AFTERMATH: WHAT JOHN BUILT

John E. Moye returned to Washington County, Georgia, in spring 1865. He had no pension, no land grants, and no recognition beyond a parole slip identifying him as a former Confederate soldier. Records show he farmed, raised seven children who survived to adulthood, and lived quietly in the community. He died in 1906 in Dublin, Georgia (then Laurens County), at age eighty-one, having lived long enough to see his great-grandchildren and to witness the South rebuilt under terms he'd never have chosen.

His gravestone lists his service: "Pvt. Co. G, 57th Ga. Inf., C.S.A." No mention of Champion Hill, Vicksburg, or Kennesaw Mountain. Just the bare fact of service.

But the legacy existed in less visible ways: in children who knew how to manage a farm through lean years, in neighbors who turned to him to mediate disputes, and in the habit of competence under pressure that he carried all his life.

His grandson, Robert George Washington Moye, born in 1852, would carry some of that competence forward—though not into war. Robert moved the family to Florida in the 1880s, where he raised his own children far from the battlefields that had consumed his grandfather's youth. The Moye family's military service would skip a generation—Robert's son, James Morris Moye, was too young for the Spanish-American War and too old for World War I—before resuming in the next century.

But the lesson John learned at Champion Hill endured: When the captain falls, someone has to lead. And if you've been paying attention, if you understand the mission, and if you're competent enough that people trust you—then that someone might be you.

## THE BENCH STRENGTH IMPERATIVE

The business term for what the 57th Georgia developed is "bench strength"—the depth of leadership capability below the starting lineup. Sports teams talk about it constantly: Can the backup quarterback run the offense? Can the bench players maintain intensity? Companies talk about it less but need it more.

A 2023 survey of tech companies that survived mass layoffs with minimal disruption found a common factor: They'd invested in leadership development at every level—not just executive coaching or management training for directors, but also programs that taught senior engineers how to make strategic decisions, that gave product managers exposure to P&L discussions, and that rotated individual contributors through cross-functional projects so they understood how the business worked beyond their immediate team.

These companies had built what one CEO called "the next-person-up culture"—the assumption that if anyone left, someone else could step into the role because they'd been learning by watching and participating. This isn't a matter of titles or org charts. It's a matter of distributing context so widely that no single person's departure destroys capability.

The companies that struggled after layoffs were those where knowledge lived in individuals rather than in systems. The critical codebase existed only in one engineer's head. The vendor relationships were managed by a single procurement specialist. The customer's needs were understood by a lone account manager. When layoffs removed these people, the knowledge vanished with them.

Building bench strength requires intentional effort:

**Cross-Training and Rotation:** Move people between roles so they understand adjacent functions. The engineer who spends a quarter shadowing the product manager learns how prioritization works. The product manager who sits with the sales team learns what customers actually care about. The sales rep who reviews the cost structure learns why certain features are expensive to build. This doesn't mean everyone does everything—specialization still matters—but it means everyone understands the context in which their specialization operates.

**Transparent Decision-Making:** When leaders make decisions, explain the reasoning. Don't just announce, "We're deprioritizing Feature X." Say, "We're deprioritizing Feature X because Customer Segment A generates 60 percent of revenue and they've requested Feature Y instead, and our capacity constraint means we can't do both." The junior person listening learns not just what decision was made but also how to make similar decisions.

**Small Bets on Emerging Leaders:** Give people authority before they're "ready." Let the senior engineer run a project planning meeting. Let the junior product manager negotiate a small vendor contract. Create opportunities for people to fail small so they don't fail big when it actually matters. Confederate regiments made sergeants out of corporals not by promoting them formally but by giving them command of a squad and seeing if they could handle it. Modern organizations can do the same with far lower stakes.

**Document Intent, Not Just Process:** Process documentation tells you how to do something. Intent documentation tells you why. "Our release process requires sign-off from QA and Product" is process. "We require QA and Product sign-off because 90 percent of customer churn comes from bugs in the first week post-release, so we optimize for stability over speed" is intent. When the person who built the process leaves, the process remains but becomes cargo cult if no one knows why it exists. Documenting intent lets the next person adapt the process intelligently when circumstances change.

## CONCLUSION: THE INHERITANCE

What John E. Moye carried from the war, beyond whatever physical scars he bore, was a lesson about leadership that transcends causes and flags: Organizations survive crisis when ordinary people can step into extraordinary roles.

The 57th Georgia Infantry survived three years of war not because it had exceptional officers—it had average ones who kept getting killed—but because it had ordinary men who'd learned to function as leaders when necessity demanded. They hadn't wanted the responsibility. Most would have preferred to follow orders and let someone else make the hard calls. But war doesn't consult preferences.

Modern organizations face less deadly crises, but the principle holds. Layoffs, acquisitions, leadership transitions, and market disruptions—these are the Champion Hills of business. Companies survive them when leadership capability exists at every level, when the mission is understood by everyone, and when people are empowered to step up before they're officially asked to.

If your organization can function only when specific people are present, you don't have an organization—you have a house of cards. If it can function when those people are gone because others understand the mission and have the judgment to act, you've built something resilient.

The lesson floats along with the current of history: When the captain falls, the mission doesn't stop. It just waits to see if anyone else knows what to do. The 57th Georgia knew. And because they knew, some of them survived, their families continued, and the lesson was passed down through generations that would face their own moments when everything fell apart and someone had to step into the gap.

That's the inheritance. Not glory, but competence. Not heroism, but readiness. Not rank, but the quiet knowledge that when the moment comes, you'll know what to do because you've been watching how decisions get made, internalizing the logic, and preparing to act when circumstances demand it.

And in the end, that's all any leader can pass on: the example of how to keep going when everything else has stopped.

# CHAPTER 10

# *The Liaison*

## THREE THOUSAND METERS FROM THE WIRE

THE ORPHANAGE SAT IN a clearing three thousand meters southeast of Camp Bearcat's perimeter, close enough that Captain James Moye III could hear the afternoon artillery drills from the firebase. At the same time, he stood in ankle-deep mud talking to a Buddhist monk who wanted nothing to do with him.

It was February 1968, two weeks after the Tet Offensive had torn through Saigon and the surrounding provinces. The orphanage—calling it that was generous; it was seven palm-thatch huts with leaky roofs and dirt floors—housed seventy-two children and fourteen monks and nuns who'd fled the fighting in Bien Hoa. The children ranged from infants to teenagers, all of them war orphans, all of them living on a diet of rice so thin you could see through it and water drawn from a creek that ran brown with runoff from the surrounding paddies.

Jim Moye was twenty-seven years old that winter, a captain in the U.S. Army assigned as civil affairs officer to the 2nd Battalion, 47th Infantry Regiment, 9th Infantry Division. His official title meant he was supposed to coordinate with local Vietnamese officials, manage refugee relief, and generally be the friendly American face in a war zone where friendly faces were scarce. What it actually meant was that he spent most of his time trying to solve problems no one else wanted to touch.

Jim had learned about the orphanage from a South Vietnamese Army captain who mentioned it in passing: "There are monks out there trying to feed children. They have nothing." Jim had driven out the next morning with a sergeant and an interpreter, expecting to find maybe a dozen kids in need of a few bags of rice. What he found instead were seventy-two children and a monk named Nguyễn Văn Siu, who looked at Jim's uniform and said, through the interpreter, "We do not want your money. Money brings evil."

Jim had dealt with enough Vietnamese officials to know the script: Americans arrive with supplies, local authorities demand payment or kickbacks, and half the aid disappears into black markets. But Mr. Siu wasn't asking for money. He was asking for nothing, which was somehow more challenging. Jim stood in the mud, watching children with distended bellies and infections that should have been treated weeks before, and he tried to figure out what the hell he could do about it that wouldn't make things worse.

"What do you need?" Jim asked.

Mr. Siu, through the interpreter: "We need to build. The roofs leak. The children are sick. We have no medicine. We have no clean water."

"If I bring you supplies—lumber, medicine, tools—will you accept them?"

The monk paused. He looked at Jim carefully, the kind of look that tries to read whether a man's intentions match his words. Then: "If you bring what you say, we will build. But no money. Money corrupts."

Jim nodded. "No money."

He drove back to Camp Bearcat and started making calls.

## THE CHALLENGE OF LEADING WITHOUT AUTHORITY

What Jim faced at the orphanage was the same challenge he'd face for his entire tour in Vietnam: How do you lead when you have no authority?

He couldn't order the monks to accept American help. He couldn't command the well-drilling team to work on their day off. He couldn't direct the supply sergeant to give him lumber—the sergeant reported to a different chain of command. Jim had responsibility without authority, mission without command.

This is the fundamental challenge of coalition leadership, whether you're coordinating with Buddhist monks, allied forces, or—as we'll see—pharmaceutical partners developing drugs. You can't give orders. You can't mandate compliance. You can only build relationships, articulate shared purpose, and prove yourself competent enough that people want to work with you.

Over the next three months, Jim coordinated the orphanage project without ever filing official paperwork or requesting permission. He scrounged scrap lumber from engineering units. He diverted medical supplies from stocks designated for civilian relief. He convinced a well-drilling team to spend a weekend digging a well at the orphanage instead of taking a scheduled R&R.

When a *Stars and Stripes* reporter visited in April and asked why help an orphanage when there's a war to fight, Jim's answer became the sentence his family would remember decades later: "These are still people. That's what we're here for."

## SEPTEMBER 1967: THE QUEEN'S COBRAS ARRIVE

Six months before the orphanage project, on September 21, 1967, the first Thai combat forces deployed to South Vietnam. They called themselves the "Queen's Cobras"—a regimental-sized force of 2,200 men sent by the Thai government as part of the broader coalition supporting South Vietnam. They landed at Saigon's port, trucked north to Camp Bearcat in Bien Hoa Province, and set up alongside the U.S. 9th Infantry Division.

The problem was immediate and obvious: How do you coordinate combat operations with forces you don't command?

The Thai soldiers answered to Bangkok, not MACV (Military Assistance Command, Vietnam). They had their own officers, their own rules of engagement, and their own procedures. American commanders couldn't order them to patrol a specific sector, couldn't direct them to support American operations, and couldn't even reliably communicate with them—most Thai officers spoke limited English, and almost no American officers spoke Thai.

Jim Moye, as the civil affairs officer and, by virtue of speaking passable Vietnamese (which helped with interpreters), became the informal liaison between his battalion and the Thai forces. It wasn't an official assignment at first—no orders cut, no additional pay, and no change in duty roster. It was just the practical reality that someone needed to talk to the Thais, and Jim was the one who could manage it.

The first challenge was language. They communicated through interpreters fluent in French, English, Vietnamese, and Thai, with varying degrees of fluency. Conversations that should have taken five minutes stretched to thirty.

The second challenge was culture. Thai soldiers came from a Buddhist military tradition that emphasized hierarchy, respect for elders, and ritualistic formality. American soldiers valued directness, informality, and efficiency. Misunderstandings were constant.

Jim learned to navigate both. He showed up at the Thai headquarters with proper formality, accepted tea even on a 95-degree day, and waited through the pleasantries before discussing operations. More importantly, he made himself trustworthy. When he said artillery support would be available, it was available. When he promised medical evacuation for Thai casualties, the helicopters came. Competence built trust, and trust enabled coordination.

## MARCH 1968: COALITION UNDER FIRE

The Tet Offensive tested whether the coalition would hold under pressure.

On the night of January 31, 1968, Viet Cong forces launched simultaneous attacks across South Vietnam. In III Corps, they hit Saigon, the Long Binh logistics complex, and Bien Hoa Airbase. Jim's battalion, the 2/47th Infantry, was ordered to reinforce Long Binh, where Viet Cong sappers had penetrated the perimeter and blown up ammunition bunkers in spectacular explosions.

The battalion raced north in their M113 armored personnel carriers and spent the night clearing Viet Cong from the depot. By dawn, they were redirected into Saigon itself to help dislodge enemy forces from the Cholon district. For mechanized infantry trained to fight in open terrain, urban combat was alien and terrifying.

Jim, coordinating between his battalion and the Thai forces still holding Bearcat, spent seventy-two hours with almost no sleep, managing casualty evacuations and supply runs. At the same time, his unit fought house-to-house battles they'd never trained for.

The Thais, meanwhile, held Bearcat and conducted aggressive patrols to prevent follow-on attacks. They didn't panic. They didn't demand American forces abandon Saigon to protect Bearcat. They understood the mission—defend the approaches to Saigon—and they executed it.

What struck Jim, when he had time to reflect, was how the coalition held together under pressure. American commanders trusted the Thais to keep their sector without micromanaging. Thai commanders coordinated their operations with American units without requiring detailed orders. The system worked because everyone understood the shared purpose: Stop the Viet Cong from consolidating gains. This wasn't command-and-control in the traditional sense. It was distributed authority aligned around common objectives. Each unit made tactical decisions independently, but those decisions served the same strategic goal.

By mid-March, MACV created the Royal Thai Army Volunteer Force—Special Liaison Section (RTAVF-SLS)—a dedicated unit of eighteen U.S. personnel assigned to coordinate all support for Thai operations, including artillery, airstrikes, medical evacuation, logistics, and intelligence. Jim worked closely with the SLS team, functioning as the on-the-ground coordinator at Bearcat.

The SLS model was elegant in its simplicity: Embed American specialists with Thai units to coordinate support. SLS personnel lived at Thai firebases, ate Thai rations, and endured the same incoming mortar fire. They had no authority over Thai soldiers. They couldn't give orders, couldn't discipline, and couldn't make tactical decisions. What they could do was make the Thais more effective by connecting them to American firepower and logistics.

When a Thai patrol made contact with Viet Cong forces and needed air support, they'd radio their firebase. The Thai commander would assess whether his own artillery could handle it. If he needed American support, he'd tell the SLS radio operator, who would relay the request through U.S. channels. Within minutes, American F-4 Phantoms would be overhead.

The system worked because both sides invested in it. Thai commanders learned to communicate grid coordinates in American format. American forward observers learned to recognize Thai call signs. SLS personnel earned credibility not through rank but through competence and shared risk.

## THE ORPHANAGE REVISITED: PURPOSE OVER PROTOCOL

The orphanage project ran parallel to all of this—patrols, firefights, and the daily grind of base security. Throughout the spring of 1968, Jim continued to coordinate support for Mr.

Siu's monks. He scrounged lumber from engineering units. He diverted medical supplies. He convinced the well-drilling team to spend a weekend digging the well.

None of this was officially sanctioned. Jim didn't file requisition forms or get approval from headquarters. He just did it, operating on the principle that the purpose—helping seventy-two starving children—justified bending the rules.

This is leadership through purpose. Jim couldn't order the supply sergeant to give him lumber. He couldn't order the well-drilling team to work on their day off. But once everyone understood the purpose, cooperation became voluntary and enthusiastic. People want to do good. Leaders provide the opportunity and the justification.

Three months later, when the *Stars and Stripes* reporter visited, he found new buildings with watertight roofs, a seventy-two-foot well producing clean water, a small dispensary stocked with antibiotics and bandages, and children who were, for the first time in months, gaining weight.

The orphanage project also earned Jim credibility with the local Vietnamese population. Word spread that the Americans at Bearcat weren't there just to kill Viet Cong but also to help civilians. This wasn't strategic psychological operations. It was basic decency. But it had strategic effects: Villagers who trusted Jim were more likely to share intelligence about Viet Cong movements.

Coalition-building, whether between armies or between soldiers and civilians, operates on the same principle: Shared purpose creates collaboration that command structures cannot.

## THE MODERN PARALLEL: DRUG DEVELOPMENT AS COALITION WARFARE

In 2003, a small biotechnology company called Genentech was working on Avastin, a cancer drug that would eventually become one of the best-selling oncology treatments in history. But getting it from lab to FDA approval required coordinating dozens of organizations Genentech didn't control: academic researchers, clinical trial sites, contract manufacturers, regulatory agencies, hospitals, and thousands of patients.

The challenge was structurally identical to what Jim Moye faced at Camp Bearcat: How do you coordinate organizations when you have no authority to command them?

Consider the ecosystem required to develop a single drug. Academic researchers at universities discover biological mechanisms—they don't work for pharma companies; NIH grants fund them, and they publish for tenure. Physicians at hospitals run clinical trials, they're not employees; they're faculty members whose primary obligation is to patients, not trial sponsors. Contract research organizations manage trials—they're independent companies working for multiple clients simultaneously. The FDA evaluates safety and efficacy—it's a government agency with a mandate of its own.

That's at minimum a half-dozen stakeholder categories, each with different organizational structures, incentives, and bosses. Getting them to work together isn't a matter of command authority. It's a matter of aligning around shared purpose and coordinating through clear intent.

For Avastin, the shared purpose was straightforward: Deliver an effective cancer treatment to patients who desperately need it. That purpose aligned incentives across the ecosystem. Universities wanted the prestige of breakthrough discoveries. Genentech wanted FDA approval. Physicians wanted treatment options for their patients. Patients wanted to survive. Regulators wanted safe, effective therapies on the market.

But alignment around purpose doesn't eliminate coordination challenges. Genentech functioned as a coordinator, not a commander—precisely like the SLS at Camp Bearcat. They couldn't order academic researchers to prioritize their molecule; they could make collaboration attractive only through funding and shared credit. They couldn't force physicians to enroll patients; they could only design trials that physicians believed would benefit their patients.

What set Genentech apart from competitors was its commitment to an organizational culture of cross-functional collaboration. They organized teams around diseases, not departments. The oncology team included researchers, clinicians, regulatory specialists, manufacturing experts, and commercial leads—all in constant communication. When clinical trials hit a snag, the entire team worked together to solve the problem.

They also engaged with the FDA early and often, not treating the relationship as adversarial but as a partnership toward shared purpose. The FDA doesn't work for pharma companies, but they share the goal of getting safe, effective drugs approved. Treating regulators as coalition partners rather than obstacles accelerated the approval process.

Avastin received FDA approval in 2004 and became a multi-billion-dollar drug that extended the lives of hundreds of thousands of patients. The success wasn't primarily scientific; plenty of promising molecules failed in development. It was organizational: Genentech's ability to coordinate a complex ecosystem toward a shared purpose.

The contrast is instructive. For every Genentech, dozens of biotech companies fail despite good science because they can't coordinate the ecosystem. They treat partners as vendors, regulators as obstacles, and physicians as service providers rather than collaborators. They issue purchase orders instead of building relationships. And their drugs die in development, not because the science was wrong, but because the coalition never formed.

## THE LEADERSHIP FRAMEWORK

The challenge of leading without authority runs through Jim Moye's entire Vietnam service. He couldn't order the Thai colonel to conduct joint patrols. He couldn't command the Buddhist monks to accept American help. He couldn't direct the well-drilling team to work on their day off.

What he could do was:

**Articulate shared purpose.** Every coalition Jim built started with identifying common ground. With the Thais: Defend Bien Hoa Province from attack. With the monks: Protect the children. With the well-drilling team: Provide clean water to people who need it. Purpose doesn't mean everyone has identical motivations—the Thais were there for pay, the Americans for policy objectives, and the monks for humanitarian reasons—but it means their immediate objectives overlap enough to enable cooperation.

**Demonstrate competence.** Trust in coalitions is earned through reliability. Jim's credibility with Thai forces came from delivering on promises: When he said artillery would support their patrol, it did. When he arranged a medevac for casualties, the helicopters arrived. That reliability meant Thai commanders would call Jim when they needed help rather than try to solve problems alone.

**Share risk.** The SLS operators who lived at Thai firebases, who endured the same mortar attacks and ate the same rations, earned credibility that staff officers at headquarters never could. Shared risk creates bonds that shared command structures don't. In modern organizations, this translates into leaders who understand the work their teams do and who've "been there" in some meaningful way.

**Respect autonomy.** Jim never tried to tell the Thai colonel how to run his regiment or the Buddhist monk how to organize the orphanage. He offered support, provided resources, and let the people closest to the work make tactical decisions. Coalition leadership requires resisting the instinct to take over or to "help" by telling people how to do their jobs better.

**Solve problems faster than bureaucracy.** The orphanage project succeeded because Jim didn't file the paperwork to request permission. He scrounged supplies, organized volunteers, and solved problems using informal networks. If he'd followed proper channels, the children would have starved while he waited for authorization.

## THE GATE STANDS OPEN

Jim Moye returned home from Vietnam in late 1968. His wife had given birth to their second son while he was deployed, during Tet, while explosions echoed across Saigon. He'd missed the first months of his son's life because he'd been coordinating with Thai forces, arranging lumber deliveries to Buddhist monks, and scrounging ammunition for urban firefights he'd never trained for.

The war would continue for another seven years. The Thais would eventually withdraw in 1971 as U.S. policy shifted. The orphanage, according to later reports, survived the war and became a permanent institution, the well still producing water decades later. Most of the operational records were archived, and the lessons—about coalition warfare, leadership without authority, and the importance of shared purpose—were forgotten or reduced to footnotes.

But the lessons remain valid. Organizations today are coalitions more than hierarchies. The product manager coordinating with engineering, design, sales, and support doesn't command those teams; she coordinates them. The CEO negotiating with vendors, partners, investors, and regulators doesn't give orders; he builds relationships. The project lead assembling a cross-functional team doesn't have direct reports; he has collaborators with their own bosses and priorities.

Success in these environments requires the same skills Jim Moye developed at Camp Bearcat: Articulate a shared purpose, demonstrate competence, share risk, respect autonomy, and solve problems faster than bureaucracy. It requires treating partners as partners, not subordinates. It requires the humility to admit you can't command and the confidence to believe you can still lead.

"These are still people," Jim told the *Stars and Stripes* reporter. "That's what we're here for." Not for dominance, not for control, but for cooperation in service of something larger than any single organization could achieve alone.

That's coalition warfare. That's leadership without authority. That's how you build something that endures after the shooting stops and the soldiers go home, and all that's left is the well still pulling clean water from the earth.

# CHAPTER 11

## *The Empowered Crew*

### MIDNIGHT OVER THE MEDITERRANEAN

THE P-3C ORION BANKED left over the black water south of Crete, its four turboprop engines humming at cruise power. It was April 6, 2003, and I sat at the tactical coordinator station behind the flight deck, staring at a display showing empty sea and trying to remember the last time I'd slept more than four hours.

Combat Air Crew Four—CAC-4—had been flying missions over Iraq for two weeks straight, providing Intelligence, Surveillance, and Reconnaissance (ISR) support for ground forces. The P-3, a Cold War relic designed for anti-submarine warfare, had been hastily retrofitted with cameras and sensors to support the invasion. It was an awkward adaptation, but it was what the Navy had available.

I was twenty-six years old, a lieutenant, and the tactical coordinator for CAC-4. That meant I sat in the back of the plane and made the real-time decisions: where to search, what to photograph, which radio calls to prioritize, when to extend the mission, and when to head home. The aircraft commander, a lieutenant commander we called "G-Money," technically outranked me, but in practice, we worked as partners. He flew the airplane; I ran the mission.

Tonight, the crew was barely holding together. We'd flown six missions in eight days. The crew was exhausted, frayed, and starting not to care. The eleven people scattered throughout the aircraft each played a specialized role. Three pilots rotated flying duties. Two naval flight officers managed sensors and navigation. Three sensor operators ran radar and cameras. Two flight engineers monitored engines and fuel. One in-flight technician watched for threats out the back windows.

Each person was a specialist. The pilots couldn't operate the sensors. The sensor operators couldn't fly the plane. The flight engineers didn't make tactical decisions. This specialization creates efficiency but demands coordination. How do eleven people, scattered throughout a noisy airplane, work together as a single entity?

The answer is empowerment. Each person must be trusted to perform his role without constant supervision. I couldn't watch every display, monitor every radio, and make every decision. I had to trust that when the radar operator said, "Vehicle moving south on Route 1," the information was accurate. I had to trust that when the flight engineer said, "We need to shut down an engine to make it home," the calculation was correct.

That trust isn't automatic. It's built over months of training and missions. CAC-4 had that trust. We'd flown together for a year and trained together for six months before that. We were exhausted and scared, but we were still a crew.

Then came the call that would test whether that trust was absolute.

## THE DIVERSION

Two hours into the mission, the voice came over the radio from an Air Force AWACS controller: "Madman Five-Seven, this is Magic. We have a situation. A coalition helicopter is down near Baghdad. Search and rescue is inbound, but ETA is thirty minutes. You're the closest ISR platform. Can you divert to provide overwatch?"

I keyed the intercom to the crew: "We just got tasked with search and rescue. Helicopter down near Baghdad. They need eyes on site until the rescue birds get there."

The response was immediate and unanimous: "No."

The sensor operator: "We're not equipped for that."

The flight engineer: "Baghdad's a hot zone. We'll get shot down."

One of the pilots: "This is insane. We're a sub-hunter doing ISR. We're not combat search and rescue."

I understood the objections. They were legitimate. The P-3 had no defensive weapons and no countermeasures beyond chaff and flares. Baghdad was defended by anti-aircraft artillery and missiles. A helicopter had just been shot down there. Sending the P-3 into that environment was asking for a second aircraft down.

But there were people on the ground near Baghdad right now, American or allied aircrew, possibly injured, definitely scared, waiting for someone to find them before enemy forces did. And CAC-4 had sensors to locate a downed aircraft and cameras to identify threats. The rescue helicopters wouldn't have that situational awareness. They needed someone overhead.

I took a breath and keyed the intercom.

## THE SPEECH

"All right, listen up," I said. "I hear you. I understand the risks. Baghdad's dangerous, we're not configured for this mission, and we're all exhausted. But I need you to remember something.

"We didn't join the Navy to fly safe missions. We joined because we took an oath—to support and defend the Constitution, to follow orders, and to serve our country. Right now, our brothers and sisters are on the ground near Baghdad. They're waiting for help. And we can provide that help.

"I know we're tired. I know this is scary. But think about this: If you were the one on the ground, if your helicopter had just been shot down and you were hiding in a ditch watching enemy forces move toward you, who would you want overhead? Would you want the crew that said, 'Too dangerous; we're going home'? Or would you want the crew that said, 'We're coming—hang on'?

"We're CAC-4. We're the premier crew in this squadron. We didn't get that reputation by flying the easy missions. We got it by doing the hard things when it mattered. And tonight it matters. Those people on the ground are counting on us.

"This is what we're here for. This is the mission. Not the pre-planned targets, not the routine patrol—this. Saving lives. If we turn away now, we can tell ourselves it was the wise decision, the safe decision. But we'll know. We'll know we had a chance to help and we chose not to.

"I'm not ordering you. I can't make you want to do this. But I'm telling you why I think we should. And I'm asking you to trust me one more time."

Five seconds passed. Ten. The intercom stayed silent except for engine noise. I waited.

Finally, G-Money's voice: "Madman Five-Seven is diverting to support SAR. Magic, send us the coordinates."

One by one, the rest of the crew acknowledged: "Sensors ready." "Flight engineers ready." "Nav has the course."

I exhaled. "All right. Let's go get them."

## FOUR HOURS OVER BAGHDAD

The divert took ten minutes. The AWACS controller passed coordinates for the crash site—a field southwest of the city. As we approached, the sensor operators powered up cameras and radar.

"Got it," the camera operator said. "Wreckage visible. Looks like a Blackhawk. No movement around the crash."

I relayed the information to the AWACS and to the inbound rescue helicopters. The rescue birds were still twenty minutes out. In the meantime, our job was to map the threat environment: Where were enemy forces? Were they moving toward the crash? Were there SAM batteries that could threaten the rescue?

For the next four hours, CAC-4 orbited the crash site. The radar operator tracked vehicle movements. The camera operator identified threats. The in-flight technician watched for anti-aircraft fire—tracers arcing up from the city, falling short, but a reminder that the threat was real.

It was methodical, grinding work. Each crew member executed his role without being told. The pilots held the orbit. The sensor operators monitored screens. The flight engineers watched fuel consumption.

I orchestrated, but I didn't micromanage. I set the objective: Stay overhead until the rescue is complete. I established the constraints: Don't enter Iranian airspace, and don't drop below 15,000 feet where we're vulnerable. Then I trusted the crew to execute.

This is empowerment in its purest form: clear intent, defined constraints, and trust in distributed expertise.

## THE FUEL DECISION

Two hours in, one of the flight engineers called me: "TACCO, we've got a fuel issue. We're burning faster than planned. At current consumption, we'll hit bingo fuel in ninety minutes."

Bingo fuel: the minimum required to return to base. When you reach it, you either turn around immediately or run out of gas. Ninety minutes meant we'd have to leave before the rescue was complete. The helicopters were still inbound, at least forty minutes away.

"Options?" I asked.

"We can shut down one engine. Reduces fuel burn by about 25 percent, buys us another hour. But we lose speed and altitude performance. If we get engaged, we won't be able to maneuver as well."

I considered. Shutting down an engine was a calculated risk. The P-3 could fly on three engines, but performance degraded. If we needed to evade a missile or climb rapidly, we'd be slower and less responsive. But if we didn't shut down the engine, we'd have to leave before the rescue was complete.

"Do it," I said. "Shut down number three. We're staying until the rescue is done."

The engineers ran through the shutdown checklist. The P-3 shuddered slightly and then settled into a new rhythm. The FE adjusted the power on the remaining engines.

I made that decision without consulting squadron headquarters. There wasn't time. I assessed the situation, understood the mission, weighed the risks, and made the call. This is the commander's intent in action: the person on the scene, with the best information, makes the decision. Two hours later, the rescue was complete. The helicopters lifted off with the downed aircrew—all survived, with minor injuries—and headed south.

Except we couldn't return to Crete. The fuel calculation showed we'd come up short. We diverted to Saudi Arabia, landed at a small coalition airfield, refueled, and returned to base six hours behind schedule.

When we landed, the squadron operations officer asked for the mission summary. I ran through it: diverted to SAR, provided overwatch for four hours, shut down an engine to stay on station, and landed in Saudi Arabia to refuel.

His response: "Good work. Get some sleep. You're off the flight schedule for forty-eight hours."

That was it. No medals. Just an acknowledgment that we'd done what we were trained to do.

## THE MODERN PARALLEL: ZAPPOS

In 1999, Tony Hsieh invested in Zappos, an online shoe retailer. The challenge: Selling shoes online was considered impossible. Shoes require trying on, sizing varies, and returns would be astronomical. Hsieh's solution was radical: Empower customer service representatives—the lowest-paid employees in most companies—to create experiences so remarkable that customers would tell stories about them.

Most e-commerce companies treated customer service as a cost center to minimize: automate everything, script every interaction, and measure success by call duration. Hsieh did the opposite. He hired customer service reps in Las Vegas, trained them extensively, and then gave them something almost no customer service rep has: complete discretion to do whatever they think will create a remarkable experience.

The commander's intent Hsieh communicated was simple: We exist to deliver happiness to customers. Create an experience so good the customer tells someone about it. Act ethically and protect the brand. Everything else is your judgment.

With that framework, Zappos reps began doing things that would get employees fired at most companies:

- A rep sent flowers to a customer whose mother had just died
- A rep upgraded a customer to overnight shipping for free because the shoes were for a wedding
- A rep spent ten hours on the phone with a lonely customer who needed someone to listen
- A rep helped a customer find shoes at a competitor's website when Zappos was out of stock

Each decision costs money in the short term. But Hsieh's calculation was different: The long-term value of a customer telling a story about Zappos far exceeded the short-term cost.

Zappos grew from $1.6 million in sales in 2000 to $1 billion in 2008. Amazon acquired them in 2009 for $1.2 billion—not because they had revolutionary technology, but because they'd built a brand so strong that customers chose them over cheaper alternatives.

The brand was built one empowered customer service rep at a time.

But empowerment without competence is chaos. Zappos invested heavily in training. New hires spent four weeks learning company culture, product knowledge, and customer service philosophy. At the end of the first week, Zappos offered every new hire $2,000 to quit. About 10 percent took it. The 90 percent who stayed were more committed because they'd actively chosen to be there.

The parallel to my P-3 crew is exact. Both were specialized teams empowered to make decisions within a clear framework. Both succeeded because competence was built through training before autonomy was granted.

## HOW TO BUILD EMPOWERED TEAMS

The challenge most leaders face: They say they want empowered teams, but when given the chance to actually empower, they hesitate. Control feels safer than trust, even when control is demonstrably less effective.

Consider the typical corporate approval process: An engineer identifies a bug causing customer complaints. Fixing it requires changing shared code. The engineer files a ticket, waits for architectural review, security review, QA approval, product management sign-off, and deployment approval. Two weeks later, the fix ships. Meanwhile, customers continue to experience the bug.

This happens because someone up the hierarchy doesn't trust the person closest to the problem to make the right decision. So he inserts himself into the approval chain, ostensibly to "ensure quality" but actually because letting go feels risky.

The irony: This control creates the high risks it's designed to prevent. The bug that took two weeks to fix caused more customer damage than a bad fix would have caused.

Leaders cling to control because mistakes are visible and painful, while missed opportunities are invisible. But velocity matters. The organization that can make faster decisions will win more often than the one that makes perfect decisions slowly.

Here's how to build empowerment without chaos:

## STATE THE WHY RELENTLESSLY

Purpose must be communicated so often that it becomes boring to leadership and internalized by everyone else. At Zappos, Hsieh talked about "delivering happiness" in every meeting, every training session, and every interview. When it became a cliché, he kept saying it.

For my crew, the purpose—"support and defend, protect our fellow service members"—was drilled into us through training and reactivated by my speech when they'd temporarily lost sight of it.

The CEO is the chief repetition officer. If you're tired of saying it, you're halfway to everyone else understanding it.

## DEFINE THE NON-NEGOTIABLES

Constraints are liberating because they define boundaries within which people can operate freely. At Zappos, the constraint was "act ethically; protect the brand." Simple, sufficient.

For the P-3 crew: "Don't violate rules of engagement, don't enter Iranian airspace, and maintain minimum safe altitude." Within those limits there was enormous discretion.

What's essential: Constraints must be few, clear, and genuinely non-negotiable. If you have fifty constraints, you don't have empowerment; you have a rulebook.

## TRAIN JUDGMENT, NOT COMPLIANCE

The difference between judgment and compliance training is the difference between "What should you do in this situation?" and "What does the policy say?"

Zappos trained judgment through scenarios: "A customer's shoes for their daughter's graduation arrived in the wrong size, and graduation is tomorrow. What do you do?" There's no script. The rep has to think: What's the purpose? Make the customer happy. What outcome do we want? Customer tells a story about how Zappos saved the day. Judgment: Overnight, to a new pair, offer to let them keep the wrong-size pair.

This takes time. It requires experienced employees to coach newer ones. But the payoff is employees who can handle novel situations.

## REWARD GOOD DECISIONS, NOT JUST GOOD OUTCOMES

Sometimes good decisions lead to bad outcomes because the world is probabilistic. If you punish people for bad outcomes even when their decision-making was sound, you teach them not to take risks.

Zappos made this explicit. A rep who sent flowers to a customer who turned out to be allergic (intent good, outcome bad) wasn't punished. The company reviewed: Was the rep trying to create a remarkable experience? Yes. Did they act within the values? Yes. Did they learn for next time? Yes.

## USE AFTER-ACTION REVIEWS

After every mission, the P-3 crew conducted a debriefing: What was supposed to happen? What actually happened? Why was there a difference? What will we do differently next time? The debriefing wasn't about blame. It was about extracting lessons while memory was fresh. Organizations that master after-action reviews build institutional learning that compounds over time.

## CONCLUSION

On the morning of April 7, 2003, when my crew landed back at base after eleven hours airborne, exhausted and wired, the operations officer's response was: "Good work. Get some sleep." No medals, no formal recognition. Just an acknowledgment that we'd done what we were trained to do: execute the mission despite circumstances we hadn't planned for.

What I took from that mission was a crystallized understanding of what my family had been demonstrating for four generations: Organizations survive crises when ordinary people can step into extraordinary roles because they understand the purpose, have the skills, and are trusted to act.

George Moye II couldn't micromanage militia scattered across North Carolina. He gave them purpose and trusted them to execute. Jim Moye, coordinating with the Thai forces, couldn't command them, but he could align them through shared purpose. I couldn't order my crew to divert to Baghdad, but I could articulate why it mattered and trust them to choose.

The through-line isn't tactics or technology. It's the understanding that empowerment—distributed authority guided by shared intent—beats centralized control when the environment demands adaptation. In each case, the leader's job wasn't to make every decision, but to make it possible for others to make good decisions: by articulating purpose, by building competence, by demonstrating trust, and by clearly defining constraints.

That's how teams survive when the plan falls apart. That's how organizations adapt when circumstances change. That's how eleven people scattered throughout a noisy airplane over Baghdad can function as a single entity, making life-or-death decisions in real time.

The well still pulls water from the earth. The crew still executes the mission. The principles remain constant.

## EPILOGUE: THE RIVER RUNS CLEAR

Twenty-one years after the Baghdad mission, I sat in my office at a healthcare technology company and thought about empowerment. My team was struggling with a product launch. The engineering lead wanted to delay for more testing. The sales team wanted to ship immediately to hit quarterly targets. The product manager was caught in the middle, paralyzed by competing demands.

I called a meeting. I didn't give orders. Instead, I asked three questions:

"Why does this launch matter? Not to hit a number, but *why* does it matter to our customers, our mission, our company?"

"What outcome do we have to achieve? If we delay, what happens? If we ship with known bugs, what happens? What's the actual success condition?"

"What constraints can't we violate? What would be unacceptable—to customers, to our reputation, and to our values?"

The team talked for thirty minutes. By the end, they'd reached consensus: Ship a limited beta to friendly customers who understood it was early, gather feedback aggressively, fix critical issues in two-week sprints, and launch broadly in six weeks. It wasn't the sales team's preferred plan or the engineering lead's preferred plan, but it achieved the outcome (get customer feedback, move the product forward) without violating constraints (don't alienate customers with a broken product, but don't miss the market window entirely).

I didn't make the decision. The team made it. I just provided the framework.

That night, I thought about my great-great-great-great-great-grandfather standing in a courthouse in Pitt County, North Carolina, helping neighbors decide whether to pursue horse thieves without waiting for royal authority. I thought about my great-great-grandfather holding a trench line at Vicksburg with no officers left to give orders. I thought about my father helping Thai soldiers and Buddhist monks in a war zone where command authority meant nothing compared to shared purpose. And I thought about myself, at twenty-six years old, asking an exhausted crew to trust me one more time because lives depended on it.

The lesson, passed down like an heirloom through generations, was simple: Leadership isn't control. Leadership is clarity—clarity of purpose, clarity of outcomes, and clarity of constraints. When people know *why* they're working, *what* success looks like, and *what* lines they cannot cross, they don't need someone telling them *how* to do every task. They figure it out. And they're better at figuring it out than any leader could be, because they're closer to the work, they have specialized knowledge, and they're the ones who'll have to execute.

Empowerment isn't the absence of leadership. Its leadership matured past the need for constant intervention. It's trusting that the seeds you've planted—the purpose you've articulated, the competence you've trained, and the values you've modeled—will bear fruit even when you're not there to supervise the harvest.

The river runs clear when it runs free, carrying purpose downstream to every tributary, every creek, and every spring that feeds the current. George Moye understood that in 1775. John Moye proved it in 1863. Jim Moye demonstrated it in 1968. I lived it in 2003.

And the lesson flows on to whoever's willing to trust that the people closest to the work know how to do the work, if only you'll give them the clarity to act and the freedom to choose.

That's empowerment. That's leadership. That's the inheritance that matters more than land or titles or medals: the knowledge that when everything falls apart and no one's there to give orders, someone will step forward because they understand the mission, they trust their competence, and they know their team will back them up.

The gate stands open by your hand. The mission continues because you chose to act. The river remembers those who trusted it to flow.

# CHAPTER 12

## *What I Learned*

It started with my son's curiosity about our family name.

HE WAS SIXTEEN WHEN he began asking where the Moyes came from. We had never really explored our family tree. I knew my dad's family was from Florida, that there were relatives scattered throughout the Deep South, that there were rumors about our origins, but that was about it. So we did a DNA test and started building out the family tree.

What I found stopped me cold.

Tracing the Moye name back from my father, James Morris Moye III, I found a direct line of military service spanning the entire history of American warfare. My dad had served 25 years in the Army, including tours in Vietnam during some of the most intense fighting of that conflict. His great-great-grandfather, John E. Moye, had served four years in the Confederate Army, fighting at Vicksburg. And John's ancestors, George Moye II and his son George Moye III, had both served in the North Carolina militia from 1775 to 1781.

I sat with that for a long time. Four generations. Four wars. A family line present at the founding of this nation and at every defining military moment in the centuries that followed. I felt something I hadn't anticipated: an overwhelming sense of pride. Not the chest-puffing kind, but something quieter and deeper, the feeling of standing at thc cnd of a very long line of people who, when their moment came, showed up. Who signed their names when signing meant risk. Who held the line when holding it was costly. Who served not because it was easy, but because they believed it mattered.

When I posted about this discovery on LinkedIn, it became the most popular thing I had ever written. More engagement than anything I had shared about AI, healthcare, or technology. People were fascinated by this continuous thread of service, a family line present at the founding of our nation and at every defining moment of American military history since.

But what struck me most was not the historical coincidence. It was a realization that arrived quietly and then would not let go: every one of these men had faced the same fundamental leadership challenges I had faced that night over Baghdad. Different wars. Different weapons. Different enemies. The same question underneath all of it: How do you lead people through uncertainty, danger, and seemingly impossible odds?

That question is what turned a genealogy project into this book. And researching the answer changed the way I understand my own life.

## WHAT I FOUND WHEN I LOOKED BACK

When you trace your family's military heritage with the intention of writing a leadership book, you expect to find inspiring moments. You don't expect to find yourself. But that is what kept happening as I dug deeper into each generation.

I should say plainly: the pride I felt early in this research never left. If anything, it deepened as I moved past the names on a family tree and into the actual texture of what these men lived through. George Moye II was in his late fifties, gout in his ankles, when he marched to Guilford Courthouse. He could have stayed home. Nobody would have faulted him. He marched anyway. John E. Moye, despite fighting on the wrong side of history (something we should never forget), survived Vicksburg, survived being taken prisoner, survived the long retreat through Georgia, and kept his men functional through all of it. My father built a well for seventy-two orphaned children in the middle of a war zone because he looked across the wire and saw human beings who needed help. These were not men chasing glory. They were men who understood what was being asked of them and answered.

Knowing that this is where I come from changed something in me. It raised the standard I hold myself to. If they could do what they did under those conditions, the least I can do is lead with the same clarity of purpose in the far more comfortable arenas I operate in.

Beyond the pride, the research kept holding up a mirror.

Captain George Moye II, drilling his militia on the banks of the Tar River, asking neighbors to risk their farms and their lives against the most powerful military force on earth: I recognized that problem immediately. I had stood in front of an exhausted crew over the skies of Baghdad, trying to explain why a mission they did not ask for and did not want still mattered. He did not motivate his men with grand talk of democracy or natural rights. He talked about the right of Pitt to govern Pitt, something concrete, personal, and specific enough to act on. That night over Baghdad, I did not move my crew with strategic objectives either. I asked them to imagine being the ones on the ground, waiting for help. Purpose must be both larger than the individual and immediate enough to feel real. Captain Moye understood that. I learned it the hard way at 25,000 feet.

John E. Moye holding his regiment together through the long middle of the Civil War, through the grinding years between the dramatic battles, when men still had to tend their farms and bury their friends and find reasons to keep fighting: I recognized that too. It is not the dramatic moments that break leaders and teams. It is the long middle. The quarter when the numbers are not moving. The initiative everyone believed in six months ago that now feels like a burden. At Paige AI, we went through multiple rounds of layoffs, FDA delays, and healthcare systems that committed to pilots and then backed out. Some people left because they lost faith. Sustaining commitment through that kind of attrition is a different skill than igniting it at the beginning, and it is the skill that actually determines whether you survive.

My father, Jim Moye, coordinating with Thai forces he could not command, building relationships across cultural and organizational gaps, scrounging lumber for an orphanage nobody had assigned him to build: I grew up in the household of a man who operated that way. He just called it doing the job. But studying his service carefully, documenting the choices he made and the principles those choices reflected, I understood for the first time that what I had absorbed from watching him was a methodology. A coherent, transferable approach to leadership that had been shaped by everyone who carried this name before him.

It is also critical to state, that while the leadership lessons in this book emerge from the men of my family who endured combat across multiple wars, an equally powerful book could be written about the women who made that service possible. Military spouses embody a form of leadership that receives no medals and few public acknowledgments. The leadership of resilience, adaptability, and unwavering purpose under conditions they did not choose. Elizabeth Gardner Moye kept the family and homestead moving throughout the Revolutionary War while her husband and eldest son fought for a nation that did not yet exist. Mary A. Moye held the farm together for four years while John E. Moye was gone. My saint of a mother, Kathy Moye, navigated countless deployments and permanent changes of station, raising me and my two brothers while never knowing when our dad would return—including an entire year while he served in Korea. My wife Leanne supported me through two combat deployments and the often-harder transition to civilian life. Military spouses do not wear the uniform, but they carry a weight that few outside this world will ever understand. Without them, there would be no military, and the world cannot thank them enough.

All of this research did not just give me history. It gave me a mirror, and a standard, and a responsibility.

## HOW THIS RESEARCH CHANGED THE WAY I LEAD

I have led organizations for most of my adult life: military crews, technology startups, healthcare companies. I thought I understood the fundamentals. Writing this book showed me how much I had been doing by instinct without being able to name it, and how much I had been getting wrong while thinking I had it right.

The biggest shift was in how I think about purpose, specifically the difference between purpose as content and purpose as infrastructure.

Before this research, I treated purpose primarily as something you communicate. You craft a mission statement, you give the inspiring talk at the all-hands meeting, you connect people to the bigger picture. That matters. But it is not sufficient.

When I became CEO of Paige AI, I scheduled thirty-minute one-on-one meetings with every person in the company, about 150 people at the time. Engineers, data scientists, business development staff, administrative assistants, everyone. I asked each person the same questions: Why did you join? What gets you excited about coming to work? What are you trying to accomplish in your career? What drives you?

What I found was that people were motivated by very different things. Some were there because cancer had touched their families and they wanted to fight it. Some were there because the technical challenges of building AI that could analyze gigapixel medical images were genuinely novel and hard. Some were there because it was a good job with interesting colleagues and real career growth. All of those purposes were real, and all of them were valuable.

The mistake I had made earlier in my career, and that I see leaders make constantly, is assuming that a compelling organizational mission will mean the same thing to everyone. You give the inspirational speech about saving lives or disrupting the industry, and the people whose personal purposes naturally align with that mission get fired up, while everyone else quietly feels like they do not really belong. That is not how teams actually work. The crew that flew the Baghdad mission that night was motivated by seven or eight different things: loyalty to each other, commitment to the oath, simple professionalism, not wanting to be the one who let someone die. I did not give them one unified purpose. I reminded each of them of the purpose they already held, and showed them how this mission connected to it.

Captain Moye had understood this two hundred years before me. He did not ask his militia to be revolutionaries. He asked them to protect the right of their community to govern itself. He met them where they were.

The second shift was in how I think about honest feedback, specifically about my own resistance to it.

Studying Nathanael Greene's Southern Campaign was uncomfortable in a specific way. Greene won not because he was smarter than Cornwallis, but because he was more honest with himself about what his forces could and could not do. He did not rationalize the retreat at Guilford Courthouse. He incorporated what the evidence was telling him and acted on it. The same pattern showed up in every successful pivot I examined: Instagram, Slack, Opsware. In each case, the leaders were more afraid of being wrong and not knowing it than of being wrong and admitting it.

I have not always been that kind of leader. I have defended strategies I had already decided on. I have treated disconfirming data as a problem to be managed rather than information to be used. Writing this book gave me a clearer name for what I was doing and a clearer picture of what it costs.

The third and most personal shift was in how I understand my father.

I knew Jim Moye had served in Vietnam. I knew he had received commendations, including the Bronze Star, and done something involving Thai forces and an orphanage. But I did not know the details until I researched this book. What I found was that my father had been practicing, in a war zone, a model of leadership that most organizations never achieve in peacetime. He was coordinating people he could not command, in service of a mission that was not formally assigned to him, building trust across cultural gaps by delivering on every promise he made. When he said artillery would support a Thai patrol, it did. When he arranged a medevac, the helicopter came. He told a Stars and Stripes reporter that the children at that orphanage were "still human beings" who deserved help, not because it was in his orders, but because he understood the mission broadly enough to know that it was.

The well he arranged to have dug was still producing clean water decades after he came home.

I had spent my whole career trying to lead the way he did without fully understanding what he was doing. This book helped me understand it. And understanding it made me proud of him in a way that went beyond the ordinary pride a son feels for his father. He had carried something forward, quietly and without fanfare, that had been passed to him by men he never met. And now I carry it too.

## WHY THESE PRINCIPLES WILL STILL WORK WHEN EVERYTHING ELSE CHANGES

I want to speak directly to the reasonable skeptic, because I have been one.

When someone tells me that lessons from eighteenth-century warfare apply to modern business, my instinct is to be suspicious. Technology changes the game, and so does scale, communication speed, organizational complexity, and a workforce that has options and knows it. The conditions of 1775 have almost nothing in common with the conditions of 2026. Why should the leadership framework be the same?

Here is the honest answer: the surface conditions change completely and constantly. The underlying human problem does not change at all.

Leadership exists because organizations face challenges that no single person can solve alone, under conditions of uncertainty, with people who have their own fears, loyalties, limited information, and competing obligations. That description fit Captain Moye asking farmers to fight the British Empire. It fits a founder asking engineers to work for equity in a pre-revenue startup. It fits a product manager trying to align engineering, design, sales, and legal around a launch date that serves none of their individual priorities perfectly. Same problem. Different circumstances.

Purpose is timeless because meaning is not a luxury; it is a performance mechanism. When people understand why they are doing what they are doing, they can adapt intelligently when circumstances change. When they do not understand the purpose, they execute the directive literally until it stops working, then wait for new instructions. In a fast-moving environment, the gap between those two patterns is the gap between organizations that survive disruption and organizations that don't. George Moye II gave his militia a phrase they could use to make decisions in his absence. That is what purpose actually is: a decision-making tool. It worked in 1775. It worked at Paige. It will work in whatever organization you are leading right now.

Adaptive intelligence is timeless because reality is permanently uncertain and permanently willing to invalidate your assumptions. Every era believes its uncertainty is uniquely severe, and every era is right about that in the specific sense that the particular unknowns of any given moment are genuinely hard to navigate. But the response does not change. You need systems that generate honest feedback faster than your competitors, cultures that treat disconfirming data as valuable rather than threatening, and leaders who can distinguish between "this is hard and we should persist" and "this is wrong and we should pivot." Those are different problems with different solutions, and confusing them is how organizations die. It was true at Guilford Courthouse. It is true in today's AI-disrupted markets.

Empowered teams are timeless because centralized control has a ceiling that distributed intelligence does not. The person with the most authority in any organization always has less information about any specific problem than the person closest to that problem. Every approval layer between the person with the information and the person with the authority is a delay and a distortion. In stable environments, that inefficiency is survivable. In fast-moving ones, it is fatal. The leaders in this book, Captain Moye, John Moye, Jim Moye, and my crew over Baghdad, all succeeded not by controlling every decision but by building the conditions in which other people could make good decisions without them. That is what empowerment actually is: not a management philosophy or a motivational posture, but a system that works when the plan falls apart and no one is there to give orders.

These principles have been tested across two and a half centuries of American history, in conditions that varied enormously in every dimension except the fundamental challenge of getting people to work together toward something difficult. They will still work in twenty years, when the technologies we currently consider transformative are themselves being disrupted, because the humans doing the work will still be human.

## WHAT THIS MEANS FOR YOU

Twenty-one years after Baghdad, I sat in my office at a health-care technology company thinking about a product launch that had my team paralyzed. The engineering lead wanted to delay for more testing. Sales wanted to ship immediately to hit quarterly targets. The product manager was caught in the middle.

I called a meeting. I did not give orders. I asked three questions.

Why does this launch matter? Not to hit a number, but why does it matter to our customers and our mission? What outcome do we actually need to achieve? And what constraints can we not violate? What would be unacceptable to our customers, our reputation, our values?

The team talked for thirty minutes and reached a plan that neither the engineering lead nor the sales team had originally proposed: ship a limited beta to friendly customers who understood it was early, gather feedback aggressively, fix critical issues in two-week sprints, and launch broadly in six weeks. Not anyone's preferred plan. The right plan.

I did not make the decision. The team made it. I just provided the framework.

That meeting was a direct product of everything this research taught me. And that is what I want to leave you with: not the history, not the case studies, but the application. The history only matters if it changes what you do on Monday morning.

So let me be specific.

Audit your purpose. Not your mission statement, but the purpose you are actually communicating through your daily decisions, the questions you ask in meetings, the things you celebrate, and the things you let slide. Is it specific enough to be operational? Could someone on your team use it to make a good decision at midnight when you are unreachable? If not, it is a slogan. Slogans do not hold when things get hard. And just as importantly: understand that your team members carry different purposes into their work. Some are there because the mission personally moves them. Some are there because the work is technically fascinating. Some are there because it is a good career opportunity and they want to provide for their families. All of those are legitimate. Your job is not to give everyone the same purpose. It is to understand each person's purpose and help them see how it connects to the shared work.

Build honest feedback into your systems structurally, not just culturally. Captain Moye did not assume his militia was ready; he drilled them and measured readiness. Jim Moye did not assume the coalition was working; he lived with the Thai forces and found out. I did not assume my crew's morale was holding; I watched it deteriorate until I had to do something about it. You cannot adapt to a reality you are not seeing. The question to ask yourself is not whether you have a culture of feedback, but what specific mechanism ensures that bad news reaches you before it is too late to act on it.

Trust more deliberately. Most leaders underempower not because they believe in centralized control as a philosophy, but because letting go feels risky and the cost of micromanagement is invisible. It accrues slowly and does not show up on any dashboard, not in decision speed lost, not in team capability undeveloped, not in organizational intelligence never built. The leaders in this book empowered by design. They built the conditions of clear purpose, trained competence, and defined constraints that made distributed authority coherent rather than chaotic. That is the sequence. You cannot skip to the trust without doing the work that makes it warranted.

And finally: understand that leadership is long. No single chapter of the Moye family story is the whole story. There were defeats at Camden and Champion's Hill. There were years of grinding maintenance between the battles history remembers. There were missions that ended without medals or recognition, just "good work, get some sleep." What endured was not any particular victory. It was the commitment to show up, to hold purpose steady, and to pass something useful to whoever came next.

The people you lead are watching how you carry yourself when the outcome is uncertain. They are learning from you whether they realize it or not. They will carry what they learn into the next team they join, the next organization they lead, the next generation they influence. That is the actual inheritance at stake: not the victories, but the way you led when winning was not guaranteed.

Years after the Baghdad mission, the crew got together for a reunion. Someone asked the question that had been in the room all night: "Why did we really do it? Why did we keep flying those missions when we were so scared?"

The answers varied.

"I didn't want to let you guys down."

"I believed in what we were doing."

"I was young and thought I couldn't actually die."

"Honestly, I'm still not sure."

All of those answers were true. All of them were valid. We were the same crew, on the same missions, facing the same dangers, and we each had different reasons for being there. That is not a weakness. That is how teams actually work.

I think about that conversation often. I think about my son's question that started all of this. I think about my father, who never used the word leadership to describe what he was doing in Vietnam, who just called it doing the job. I think about George Moye II signing that document first, then handing the pen to his son. I think about John Moye holding a trench line with no officers left, because he understood the mission well enough to continue it without being told.

I am proud of these men. I am proud of this family. Not in a way that asks anything from anyone else or requires any particular acknowledgment, but in the quiet way that comes from knowing where you come from and understanding what it cost. They did not serve for recognition. They served because they believed it was right, and because the people beside them were counting on them, and because some things are worth more than personal comfort or safety.

That is the standard they set. Writing this book made me feel the weight of it in a way I had not before. And I mean that as a gift, not a burden. Knowing what they did does not make my own challenges smaller. It makes me more willing to meet them.

The thread that connects all of them is not a bloodline. It is a set of principles, held and practiced and passed forward, not always perfectly, not without cost, but consistently enough that they endured across two and a half centuries and found their way, eventually, to a Navy lieutenant flying over Baghdad who needed them on the worst night of his career.

Those principles are yours now too. Your war may not look anything like theirs. You may never wear a uniform, never take an oath of enlistment, never face anything more dangerous than a difficult board meeting or a company in free fall. That is fine. The battlefield is not the point. The principles forged on those battlefields are.

Every leader fights a war of some kind. The war to build something that matters before the runway runs out. The war to hold a team together when the market turns against you. The war to make the right decision with incomplete information while people are counting on you to get it right. The war to stay true to your values when expedience is whispering in your ear. The stakes are calibrated differently than Vicksburg or Guilford Courthouse or the streets of Saigon. But the human challenges underneath those stakes, how you move people toward purpose, how you adapt when your assumptions fail, how you trust others to carry the mission when you cannot be everywhere at once, are the same challenges George Moye II faced in 1775, that John Moye faced in 1863, that my father faced in 1968, and that I faced over Baghdad in 2003.

The reason these principles survive across centuries is not that war is a useful metaphor for business. It is that both war and business strip leadership down to its essentials. They remove the comfortable ambiguity that lets leaders avoid hard choices. They demand clarity, because the cost of confusion is visible and immediate. And in that stripped-down state, the same truths keep emerging: people need to know why the work matters, organizations need to learn faster than their circumstances change, and the people closest to the problem are usually the ones best equipped to solve it.

You do not need a family history like this one to lead by these principles. You do not need to have served, or to have sat at a tactical coordinator station over a war zone, or to have scrounged lumber for an orphanage nobody assigned you to build. What you need is the willingness to do the hard, quiet, daily work that these principles require: articulating purpose until it becomes part of how your team thinks, building honest feedback into your systems before reality forces the conversation, and extending trust to people you have prepared to receive it.

That work is available to every leader. The field is open. The principles are proven. Give people a purpose they can act on. Adapt before reality forces you to. Trust the people closest to the work to do the work. The gate stands open by your hand.

# *Moye Family Tree*

## Direct Paternal Line

### *Captain George Moye II to LCDR William Andrew Moye*

**1.** George II 1722 Moye Captain b: 1722 in Beaufort County, NC. d: 1800 in Pitt County, North Carolina; age: 78.
*+ Elizabeth Gardner b: 16 May 1723 in Tyrrell County, North Carolina, United States of America. d: Deceased in North Carolina, USA.*

**2.** George Moye III b: 1752 in Pitt County, North Carolina, United States of America. d: 1842 in Washington County, Georgia, United States of America; age: 90.
*+ Sarah Griffin Bradbury b: 1777 in North Carolina, United States. d: 18 Jun 1852 in Jefferson County, Georgia, USA; age: 75.*

**3.** Duran Griffin Moye b: 1799 in Pitt, North Carolina, USA. m: Bef. 1830 in Washington Co., GA. d: 1850 in Davisboro, Washington, Georgia, USA; age: 51.
*+ Mary Penelope Moye b: 1801. d: 1860; age: 59.*

**4.** John E Moye b: 01 Jun 1825 in Davisboro, Washington, Georgia. m: 30 Jan 1848 in Washington, Georgia. d: 16 Oct 1906 in Dublin, Laurens, Georgia; age: 81.
*+ Mary A Moye b: May 1826 in Washington, Georgia, United States. m: 30 Jan 1848 in Washington, Georgia, USA. d: Deceased.*

**5.** Robert George Washington Moye b: 23 April 1852 in Washington County, Georgia. m: 1876. d: 13 September 1930 in DeSoto, Florida, United States; age: 78.
+ *Sarah Elizabeth Moye b: 17 January 1857 in Wrightsville, Johnson County, Georgia, United States of America. m: 16 SEP 1875. d: 8 June 1932 in Arcadia, DeSoto County, Florida, United States of America; age: 75.*

**6.** James Morris Moye b: 5 Feb 1882 in Georgia. d: 25 January 1932 in Bostwick, Putnam County, Florida, United States of America; age: 49.
+ *Zula Elizabeth CAMPBELL Moye b: 7 January 1883 in Florida. d: 23 March 1941 in Seminole, Florida, USA; age: 58.*

**7.** James Morris Moye Jr b: 6 Oct 1906 in Port Tampa, Hillsborough, Florida. m: 1930 in Seminole, Florida. d: 25 Sep 1980 in Sanford, Seminole, Florida; age: 73.
+ *Helen Merriman Wilson b: 17 Jun 1908 in Georgia. m: 1930 in Seminole, Florida, USA. d: 9 Oct 1984 in Sanford, Seminole, Florida, USA; age: 76.*

**8.** James Morris Moye III b: 19 Apr 1940 in Cleveland, Ohio. d: 27 Aug 2008 in Lubbock, Texas, USA; age: 68.
+ *Kathryn Oth b: 09 Aug 1941 in St Louis, Missouri, USA. m: 24 Aug 1964 in El Paso.*

**9.** William Andrew Moye b: 02 Sep 1976 in Frankfurt, Frankfurt am Main, Hesse, Germany.

# *Notes*

## INTRODUCTION

**1. "April 6, 2003... P-3C Orion aircraft"** The P-3C Orion is a four-engine turboprop maritime patrol aircraft developed by Lockheed for the United States Navy. Originally designed for anti-submarine warfare during the Cold War, the aircraft was adapted for overland intelligence, surveillance, and reconnaissance (ISR) missions during Operation Iraqi Freedom. For more on the P-3's role in Iraq, see the Naval History and Heritage Command archives. The UH-60 Black Hawk was officially reported shot down on April 5, but the crew was tasked to search early morning of April 6. https://en.wikipedia.org/wiki/List_of_aviation_shoot-downs_and_accidents_during_the_Iraq_War#:~:text=(2)%20 15%20April%20%E2%80%93%20Two,whether%20it%20 is%20shot%20down.

**2. "shock-and-awe invasion began in late March"** Operation Iraqi Freedom began on March 19, 2003, with the "shock and awe" air campaign targeting Baghdad. The phrase was coined in a 1996 military doctrine paper by Harlan K. Ullman and James P. Wade.

**3. "Saddam's deck of playing cards"** The "Most Wanted Iraqi" playing cards were issued by the U.S. military in 2003 to help troops identify the most-wanted members of Saddam Hussein's government. Each card featured a photograph and name of a high-value target.

**4. "sign the Pitt Association in 1775"** The Pitt Association was signed on July 1, 1775, in Martinborough (later Greenville),

North Carolina. The original document is preserved in *The Colonial Records of North Carolina*, Volume 10, edited by William L. Saunders (Raleigh: State of North Carolina, 1886-1890). The full text is available through Documenting the American South at the University of North Carolina: https://docsouth.unc.edu/csr/index.php/document/csr10-0022

**5. "John E. Moye... 44% casualties in a single day at Champion's Hill"** The Battle of Champion Hill (also called Champion's Hill or Baker's Creek) was fought on May 16, 1863, during the Vicksburg Campaign. The 57th Georgia Infantry, in which John E. Moye served, was part of Brigadier General Alfred Cumming's Georgia Brigade. Casualty figures are drawn from regimental records in the National Archives and the National Park Service Vicksburg Campaign records.

**6. "as he told Stars and Stripes"** *Stars and Stripes* is an independent daily newspaper serving the U.S. Armed Forces community. The reference is to coverage of U.S. servicemembers' humanitarian efforts during the Vietnam War. Available at: https://www.377sps.org/stripes/april/26.pdf#:~:text=LONG%20THANH%2C%20Vietnam%20%E2%80%94%20Take,to%20help%20the%20wary%20pioneers

## CHAPTER 1: PURPOSE OVER VISION

**1. "July 1775... eighty-eight citizens of Pitt County... signed what would become known as the Pitt Association"** The proceedings of the Safety Committee in Pitt County, including the full text of the Pitt Association and the names of signatories, are preserved in *The Colonial Records of North Carolina*, Volume 10. The document begins: "We the freeholders and inhabitants of the county of Pitt and town of Martin, being deeply affected with the present alarming state of this Province and of all America..." Available at: https://docsouth.unc.edu/csr/index.php/document/csr10-0022

**2. "Pitt County had been carved out of Beaufort County only fifteen years earlier, in 1760"** Pitt County was established in 1760 from Beaufort County and named for William Pitt the Elder, the British statesman who supported colonial rights. See the Pitt County Historical Society records and the North Carolina Office of Archives and History.

**3. "George Moye II and his wife Elizabeth Gardner had married in neighboring Tyrrell County in 1743"** Marriage and land records for George Moye II are preserved in the North Carolina State Archives. Land grant records show his acquisition of property near Grindle Creek in what became Pitt County.

**4. "Committees of Safety"** Committees of Safety were local governmental bodies established throughout the American colonies in 1774-1775 to enforce the Continental Association and organize resistance to British authority. In North Carolina, these committees effectively replaced royal authority at the county level. See *The Regulators in North Carolina: A Documentary History, 1759-1776* by William Stevens Powell (Raleigh: Department of Archives and History, 1971).

**5. "Moore's Creek Bridge on February 27"** The Battle of Moore's Creek Bridge was fought on February 27, 1776, near present-day Currie in Pender County, North Carolina. Patriot forces under Colonels James Moore, Richard Caswell, and Alexander Lillington defeated approximately 1,600 Loyalists, mostly Scottish Highlanders, in a brief but decisive engagement. The battle is often called the "Lexington and Concord of the

South" because it ended royal authority in North Carolina and enabled the colony to become the first to vote for independence. The battlefield is preserved as Moores Creek National Battlefield, operated by the National Park Service. See: https://www.nps.gov/mocr/

**6. "Continental quartermasters requisitioned wagons and supplies"** The Continental Army's quartermaster operations in North Carolina are documented in the papers of Nathanael Greene and in *The Southern Campaigns of the American Revolution* by Dan L. Morrill (1992).

**7. "Captain Moye's wagon teams were conscripted by Continental General Benjamin Lincoln"** General Benjamin Lincoln commanded the Southern Department of the Continental Army from 1778 until his surrender at Charleston on May 12, 1780—the largest American surrender until the Civil War. Approximately 5,500 Continental soldiers and militia were captured.

**8. "In August, they followed Horatio Gates to Camden"** The Battle of Camden, fought on August 16, 1780, was one of the worst American defeats of the Revolutionary War. General Horatio Gates's army was routed by British forces under Lord Cornwallis. The militia broke within minutes of the British advance. See the American Battlefield Trust: https://www.battlefields.org/learn/revolutionary-war/battles/camden

**9. "General Nathanael Greene had drawn Cornwallis inland"** Major General Nathanael Greene assumed command of the Southern Department on December 2, 1780. His strategy of drawing Cornwallis away from coastal supply lines while avoiding decisive battle became known as the "Race to the Dan."

See *Nathanael Greene: A Biography of the American Revolution* by Gerald M. Carbone (2008).

**10. "On March 15, 1781, at Guilford Courthouse"** The Battle of Guilford Courthouse was fought on March 15, 1781, near present-day Greensboro, North Carolina. British forces under Lieutenant General Charles Cornwallis defeated Major General Nathanael Greene's American army, but suffered such heavy casualties (approximately 27% of their force) that Cornwallis was forced to abandon his North Carolina campaign. British statesman Charles James Fox famously said, "Another such victory would ruin the British army." The battlefield is preserved as Guilford Courthouse National Military Park. See: https://www.nps.gov/guco/ and the American Battlefield Trust: https://www.battlefields.org/learn/revolutionary-war/battles/guilford-court-house

**11. "Cornwallis claimed the field, but he'd lost roughly 25% of his force"** British casualties at Guilford Courthouse were approximately 550 killed and wounded out of roughly 2,100 engaged—about 27% of his force. American casualties were approximately 250-300. Official returns are found in Cornwallis's dispatch to Lord George Germain, dated March 17, 1781.

**12. "News of Yorktown trickled south in December 1781"** The Siege of Yorktown ended on October 19, 1781, when Cornwallis surrendered approximately 8,000 British troops to General George Washington and the Comte de Rochambeau. News reached the Carolinas in the following weeks.

**13. "North Carolina's legislature compensated veterans with land grants"** North Carolina's land grant system for Revolutionary War veterans is documented in *North Carolina Land Grants in Tennessee, 1778-1791* and other records preserved in the North

Carolina State Archives. Veterans received grants based on rank and length of service.

**14. "George III served as sheriff of Pitt County"** George Moye III's service as sheriff and other public offices are recorded in Pitt County court minutes and state records, available through the North Carolina State Archives.

## CHAPTER 2: PURPOSE THROUGH CATASTROPHE

*Civil War: Vicksburg, 1863*

**1. "The 57th Georgia Infantry Regiment"** The 57th Georgia Infantry Regiment was organized in May 1862 in Macon, Georgia, as part of the Confederate States Army. The regiment served primarily in the Western Theater of the Civil War. Regimental records are preserved in the National Archives and Records Administration (NARA) in Record Group 109: War Department Collection of Confederate Records.

**2. "Battle of Champion's Hill, May 16, 1863"** The Battle of Champion Hill (also called Champion's Hill or Baker's Creek) was the decisive engagement of the Vicksburg Campaign. Union forces under Major General Ulysses S. Grant defeated Confederate forces under Lieutenant General John C. Pemberton. The 57th Georgia, part of Brigadier General Alfred Cumming's Georgia Brigade, suffered catastrophic casualties—approximately 44% of engaged strength. The battlefield is partially preserved and interpreted by the Champion Hill Battlefield Foundation. See the National Park Service Vicksburg Campaign resources: https://www.nps.gov/vick/

**3. "Confederate General John C. Pemberton surrendered Vicksburg... July 4, 1863"** The Siege of Vicksburg lasted from May 18 to July 4, 1863. Pemberton's surrender of approximately 29,000 troops gave the Union control of the Mississippi River, effectively splitting the Confederacy. Grant's decision to parole rather than imprison the Confederate soldiers was a pragmatic choice that avoided logistical burdens. See *Vicksburg: The Campaign That Opened the Mississippi* by Michael B. Ballard (University of North Carolina Press, 2004).

**4. "Gettysburg had failed. Vicksburg had fallen."** The Battle of Gettysburg (July 1-3, 1863) and the fall of Vicksburg (July 4, 1863) occurred nearly simultaneously, marking the strategic turning point of the Civil War. The dual defeats ended Confederate hopes of foreign recognition and military initiative in both theaters.

**5. "assigned to coastal defense at Savannah"** After the Vicksburg parole and exchange, the 57th Georgia was sent to Savannah for rest and reorganization. Coastal defense duties in late 1863 reflected the Confederate strategy of consolidating depleted Western Theater units while awaiting exchange formalities.

**6. "Whitemarsh Island near Savannah... February 1864"** Whitemarsh Island, located east of Savannah in the coastal marshes, was the site of minor skirmishing as Union forces probed Confederate coastal defenses. The engagement in February 1864 was a relatively small affair compared to what the 57th Georgia had endured at Vicksburg.

**7. "Andersonville Prison"** Camp Sumter, commonly known as Andersonville Prison, operated from February 1864 to May 1865 in Sumter County, Georgia. Originally designed to hold 10,000 prisoners, at its peak the stockade held over 33,000 Union soldiers in just 26.5 acres. Approximately 13,000 prisoners died, primarily from scurvy, dysentery, and exposure. The site is now Andersonville National Historic Site, operated by the National Park Service: https://www.nps.gov/ande/

**8. "Atlanta Campaign... Kennesaw Mountain in June"** The Atlanta Campaign (May-September 1864) was Major General William T. Sherman's offensive to capture Atlanta, a critical Confederate railroad and industrial center. The Battle of Kennesaw Mountain on June 27, 1864, was a rare frontal assault ordered by Sherman against entrenched Confederate positions. Union forces suffered approximately 3,000 casualties while inflicting only 800 on defenders. Kennesaw Mountain National Battlefield Park: https://www.nps.gov/kemo/

**9. "Peachtree Creek and the Battle of Atlanta in July"** The Battle of Peachtree Creek (July 20, 1864) and the Battle of Atlanta (July 22, 1864) were desperate Confederate counterattacks ordered by General John Bell Hood after he replaced General Joseph E. Johnston. Both attacks failed to halt Sherman's advance but inflicted significant casualties on both sides.

**10. "Jonesboro in late August"** The Battle of Jonesboro (August 31 - September 1, 1864) was the final major engagement of the Atlanta Campaign. Union forces cut the Macon & Western Railroad, the last rail line supplying Atlanta, forcing Confederate evacuation of the city.

**11. "Atlanta fell on September 2, 1864"** Sherman's telegram to Washington—"Atlanta is ours, and fairly won"—marked a crucial political as well as military victory. The fall of Atlanta virtually guaranteed Abraham Lincoln's reelection in November

1864, ending Confederate hopes that a Democratic administration might negotiate peace.

**12. "General John Bell Hood... marched north into Tennessee"** Hood's Tennessee Campaign (September-December 1864) was a desperate attempt to draw Sherman away from his March to the Sea by threatening Union-held Tennessee. The campaign ended disastrously at the Battles of Franklin (November 30, 1864) and Nashville (December 15-16, 1864), where Hood's Army of Tennessee was effectively destroyed as a fighting force.

**13. "consolidated into the 1st Georgia Consolidated Infantry"** By early 1865, many Confederate regiments had been so reduced by casualties and desertion that they were administratively merged into "consolidated" units. The 1st Georgia Consolidated Infantry combined remnants of multiple shattered Georgia regiments to form a single functional unit.

**14. "surrendered at Greensboro, North Carolina... April 26, 1865"** General Joseph E. Johnston surrendered the remnants of the Army of Tennessee to Major General William T. Sherman at Bennett Place, near Durham, North Carolina, on April 26, 1865. This was the largest Confederate surrender of the war, involving approximately 89,000 troops across multiple states. The surrender terms were initially rejected by Washington as too lenient before being renegotiated.

## CHAPTER 3: MISSION FIRST, PEOPLE ALWAYS

*Vietnam War: Camp Bearcat, 1967-1968*

**1. "Camp Bearcat... twenty miles southeast of Saigon"** Camp Bearcat (also known as Camp Martin Cox) was a major U.S.

military installation in Long Thanh, Bien Hoa Province, Republic of Vietnam. It served as the main base for the 9th Infantry Division and later hosted Royal Thai Army forces. The base was located along Highway 15, approximately 20 miles southeast of Saigon.

**2. "9th Infantry Division—nicknamed the 'Old Reliables'"** The 9th Infantry Division was reactivated in February 1966 specifically for Vietnam service. The division's nickname dates to World War II service. In Vietnam, the 9th Division operated primarily in the III Corps tactical zone and the Mekong Delta. Division records are held at the National Archives and the 9th Infantry Division Association maintains historical archives.

**3. "2nd Battalion, 47th Infantry Regiment"** The 2/47th Infantry was a mechanized battalion operating M113 armored personnel carriers. The battalion saw heavy action during the Tet Offensive and throughout the Vietnam War. Unit records and after-action reports are preserved in the National Archives.

**4. "Royal Thai Army Volunteer Regiment—nicknamed the 'Queen's Cobras'—arrived in Vietnam... September 1967"** Thailand was one of several Free World Military Forces contributing troops to support South Vietnam. The Queen's Cobras were the initial Thai deployment, comprising approximately 2,200 soldiers. They were followed by the much larger Black Panther Division (approximately 11,000 troops) in mid-1968. See *Allied Participation in Vietnam* by Stanley Robert Larsen and James Lawton Collins Jr. (Department of the Army, 1975).

**5. "Operation Junction City (February–March 1967)"** Operation Junction City was the largest U.S. offensive of the Vietnam War to that point, involving over 25,000 troops searching for COSVN (Central Office for South Vietnam—the communist headquarters for southern operations) in War Zone C, Tay Ninh Province. The operation included the only U.S. combat parachute assault of the Vietnam War.

**6. "Battle near Bau Bang, March 20, 1967"** The Battle of Bau Bang II occurred on March 20, 1967, when a mechanized cavalry troop was ambushed by Viet Cong forces. Despite being outnumbered, the American unit repelled multiple assaults, inflicting heavy casualties on the attackers.

**7. "Operations Enterprise and Akron... Route 15 (nicknamed 'Thunder Road')"** These operations involved securing key transportation routes in the III Corps area. Route 15 connected Saigon to the coastal areas and was a frequent target of Viet Cong interdiction.

**8. "Mobile Riverine Force... Navy Task Force 117"** The Mobile Riverine Force was a joint Army-Navy unit operating in the Mekong Delta, combining infantry battalions with Navy river patrol boats. Task Force 117 provided the naval component, operating modified landing craft and patrol boats. See *The Riverine Force* by R.L. Schreadley (1969).

**9. "Operation Coronado VIII... 'Forest of Assassins'"** Operation Coronado VIII took place in September 1967 in the Rung Sat Special Zone, nicknamed the "Forest of Assassins"—a dense mangrove swamp area that provided cover for Viet Cong forces threatening Saigon's shipping approaches.

**10. "Tet Offensive—January 31, 1968"** The Tet Offensive was a coordinated series of surprise attacks by Viet Cong and North Vietnamese Army forces against South Vietnamese cities, towns, and military bases during the Lunar New Year (Tet) holiday ceasefire. Approximately 70,000 communist troops attacked over 100 targets simultaneously. Though a tactical defeat for the communists (who suffered enormous casualties), Tet was a strategic and psychological turning point that undermined American public support for the war. See *The Tet Offensive: A Concise History* by James H. Willbanks (Columbia University Press, 2007).

**11. "Long Binh Post"** Long Binh Post was the largest U.S. Army base in Vietnam, serving as the headquarters for II Field Force, Vietnam, and housing massive logistics and ammunition facilities. The base was attacked during Tet, with Viet Cong sappers penetrating the ammunition depot.

**12. "Cholon, the dense urban district in southern Saigon"** Cholon was Saigon's Chinatown—a densely populated commercial and residential area. Heavy fighting occurred in Cholon during both Tet and Mini-Tet as U.S. and ARVN forces cleared Viet Cong forces from urban strongpoints.

**13. "Mini-Tet... May 1968"** The May Offensive (sometimes called Mini-Tet) was a second wave of communist attacks in May 1968. Though smaller than the January offensive, Mini-Tet demonstrated continued communist capability and led to additional urban combat in Saigon.

**14. "Cau Mat hamlet and the Y-Bridge area"** The Y-Bridge was a strategic crossing point in southern Saigon's 8th District. Fighting at Cau Mat hamlet and the Y-Bridge during Mini-Tet involved intense house-to-house combat as mechanized infantry units cleared Viet Cong forces.

**15. "Mr. Nguyễn Văn Siu"** The Buddhist monk who led the orphanage near Camp Bearcat. His statement that "We don't want your money. Money brings evil" reflected the monks' preference for direct material assistance over cash donations.

**16. "Stars and Stripes interviewed him about the project"** *Stars and Stripes* is an independent daily newspaper serving the U.S. Armed Forces community since 1861. Coverage of humanitarian civic action projects was common during Vietnam as the military sought to publicize "hearts and minds" efforts.

**17. "72-foot water well dug by hand"** Hand-dug wells in Vietnam required significant labor but provided long-lasting clean water sources. The depth of 72 feet suggests the well reached a reliable aquifer below the seasonal water table.

**18. "II Field Force created a Special Liaison Section (SLS)"** The Royal Thai Army Volunteer Force – Special Liaison Section (RTAVF-SLS) was established in March 1968 to coordinate U.S. support (artillery, air support, medical evacuation, logistics) for Thai military operations. The unit comprised approximately 18 U.S. personnel embedded with Thai forces.

## CHAPTER 4: THE PURPOSE EACH PERSON CARRIES

*Iraq War: Northern Iraq, 2003*

**1. "March 26, 2003... Souda Bay Naval Air Station, Crete"** Souda Bay is a major NATO naval base on the Greek island of Crete. During Operation Iraqi Freedom, it served as a staging point for U.S. Navy P-3 Orion aircraft conducting surveillance missions over Iraq.

**2. "Combat Air Crew Four—CAC-4"** U.S. Navy P-3 Orion squadrons organize into numbered "Combat Air Crews" that train and deploy together. CAC-4 flew some of the first overland ISR missions into Iraq during the opening weeks of Operation Iraqi Freedom.

**3. "first U.S. Navy P-3 Orion crew to fly an overland Intelligence, Surveillance, and Reconnaissance mission into northern Iraq"** The P-3C Orion, designed for anti-submarine warfare, was adapted for overland ISR missions beginning in the 1990s. The aircraft's long endurance and sensor packages made it valuable for surveillance, though it lacked the self-defense capabilities of purpose-built reconnaissance aircraft. Available at: https://www.airlant.usff.navy.mil/Organization/COMPATRECONGRU/COMPATRECONWING-11/Patrol-Squadron-VP-5/Command-History/#.

**4. "Paige AI... first FDA-approved artificial intelligence algorithm for detecting prostate cancer"** Paige AI (originally Paige.AI) developed computer vision algorithms for digital pathology. Their technology analyzes digitized tissue slides to assist pathologists in cancer detection. The company spun out of research at Memorial Sloan Kettering Cancer Center. Paige was acquired by Tempus AI in 2024.

**5. "Memorial Sloan Kettering Cancer Center"** Memorial Sloan Kettering is one of the oldest and largest private cancer centers in the world, located in New York City. Founded in 1884, it is a global leader in cancer research and treatment.

## CHAPTER 5: THE CALCULATED RETREAT

*Revolutionary War: Guilford Courthouse Revisited*

**1. "Nathanael Greene's strategic concept"** Major General Nathanael Greene commanded the Continental Army's Southern Department from December 1780 until the war's end. His strategy of avoiding pitched battles while wearing down British forces through maneuver and attrition proved decisive in the Southern Campaign. See *The Road to Guilford Courthouse: The American Revolution in the Carolinas* by John Buchanan (Wiley, 1997).

**2. "Greene's battle plan at Guilford Courthouse—three lines"** Greene deliberately deployed in depth with three defensive lines: North Carolina militia in the first line, Virginia militia in the second, and Continental regulars in the third. The militia were instructed to fire and retreat rather than stand and fight British regulars—a pre-planned tactical withdrawal that preserved forces while inflicting casualties.

**3. "Charles James Fox... 'Another such victory would ruin the British army'"** Charles James Fox (1749-1806) was a leading Whig politician who opposed the war with the American colonies. His quote about Guilford Courthouse (sometimes attributed to others) captured the pyrrhic nature of British tactical victories that came at unsustainable costs.

**4. "Burbn... Instagram... October 2010"** Burbn was a location-based check-in app founded by Kevin Systrom and Mike Krieger in 2010. When the check-in features failed to gain traction, the founders noticed users engaging heavily with the photo-sharing feature. They pivoted to focus exclusively on photo sharing with filters, relaunching as Instagram in October 2010. Facebook acquired Instagram for approximately $1 billion in April 2012.

**5. "Loudcloud... Opsware... Hewlett-Packard acquired it for roughly $1.6 billion"** Loudcloud was founded by Ben Horowitz and Marc Andreessen in 1999 as a managed hosting service. After the dot-com crash devastated their customer base, Horowitz pivoted the company to enterprise software, selling the hosting business and refounding as Opsware. Hewlett-Packard acquired Opsware in 2007 for $1.65 billion. Horowitz recounted this experience in *The Hard Thing About Hard Things* (Harper Business, 2014).

**6. "Slack"** Slack Technologies was born from the failure of Glitch, an online game developed by Stewart Butterfield's company Tiny Speck. The internal communication tool built to coordinate game development proved more valuable than the game itself. Butterfield pivoted to focus on the communication tool, launching Slack in 2013. Salesforce acquired Slack in 2021 for approximately $27.7 billion.

**7. "ChatGPT launched in late November 2022"** OpenAI released ChatGPT on November 30, 2022. The application reached 100 million users within two months, making it the fastest-growing consumer application in history at that time.

**8. "transformer architecture that made ChatGPT possible"** The transformer architecture was introduced in the seminal paper "Attention Is All You Need" by Vaswani et al. at Google Brain in 2017. This architecture, particularly the attention mechanism, enabled the large language models that power modern AI assistants.

**9. "Google Bard launched in February 2023... gave a factually incorrect answer about the James Webb Space Telescope"** Google announced Bard on February 6, 2023, but a factual error in the promotional demo (incorrectly stating JWST took the first pictures of a planet outside our solar system) contributed to Alphabet's stock dropping approximately $100 billion in market value.

**10. "Microsoft... partnership with OpenAI"** Microsoft invested $1 billion in OpenAI in 2019, followed by additional investments totaling approximately $13 billion. This partnership allowed Microsoft to integrate OpenAI's technology into products like Bing, Office (Copilot), and Azure.

**11. "Notion and Linear"** Notion (founded 2016) and Linear (founded 2019) are productivity software companies that built AI features into their core products from early stages, rather than retrofitting AI onto legacy systems.

**12. "Microsoft under Steve Ballmer versus Microsoft under Satya Nadella"** Steve Ballmer served as Microsoft CEO from 2000 to 2014. Satya Nadella succeeded him and transformed the company's culture and strategy, shifting focus from Windows to cloud computing and platform-agnostic services. Microsoft's market capitalization increased from approximately $300 billion when Nadella took over to over $3 trillion by 2024.

**13. "Nadella's first act as CEO in 2014 was rewriting the mission"** Nadella replaced Microsoft's previous mission ("a computer on every desk and in every home") with "empower every person and every organization on the planet to achieve more"—a platform-agnostic statement that freed the company to pursue mobile, cloud, and AI opportunities.

**14. "IBM pivoted from tabulating machines to mainframes to PCs to enterprise services"** IBM's corporate history spans multiple technological transitions: mechanical tabulating machines (1890s-1950s), mainframe computers (1950s-present), personal computers (1980s-2000s), and enterprise services and cloud computing (1990s-present). The company's longevity reflects repeated willingness to cannibalize existing product lines.

**15. "Lou Gerstner took over a dying IBM in 1993"** Louis V. Gerstner Jr. served as IBM CEO from 1993 to 2002, transforming the company from a struggling hardware manufacturer into a services and software company. His account of the turnaround is *Who Says Elephants Can't Dance?* (Harper Business, 2002).

**16. "Kodak invented the digital camera in 1975"** Steven Sasson, a Kodak engineer, built the first digital camera prototype in 1975. Kodak leadership suppressed the technology because it threatened their film business. Kodak filed for bankruptcy in 2012.

**17. "Nokia 9000 Communicator in 1996"** The Nokia 9000 Communicator was a pioneering smartphone-like device featuring email, web browsing, and PDA functions. Despite this early innovation, Nokia failed to adapt to the touchscreen smartphone era, selling its mobile phone business to Microsoft in 2014.

**18. "Stephen Elop... 'burning platform' memo... 2010"** Nokia CEO Stephen Elop sent an internal memo in February 2011 (commonly dated to 2010 in discussions) comparing Nokia to a man on a burning oil platform who must jump into freezing water to survive, acknowledging the company's existential crisis.

## CHAPTER 6: STREET FIGHTING IN SAIGON

*Vietnam War: Tet Offensive*

**1. "The night of January 31, 1968"** The Tet Offensive began during the early morning hours of January 30-31, 1968 (timing varied by location due to time zone differences and attack coordination). The attacks coincided with the Lunar New Year, when many ARVN soldiers were on leave.

**2. "Seventy thousand Viet Cong and North Vietnamese Army troops attacked more than a hundred cities"** Estimates of communist forces involved in Tet range from 70,000 to 85,000 troops. The offensive struck 36 of 44 provincial capitals, 5 of 6 autonomous cities, 64 of 242 district capitals, and numerous military installations.

**3. "M113 armored personnel carriers"** The M113 was the most widely used armored personnel carrier of the Vietnam War. The aluminum-armored tracked vehicle provided mobility and protection for infantry in diverse terrain. Variants included command vehicles, mortar carriers, and vehicles with enhanced armament.

**4. "9th Division had killed over 1,600 Viet Cong and NVA troops"** Combat statistics from Tet must be viewed with the caveats that body counts were notoriously unreliable and subject to inflation. However, communist forces did suffer catastrophic casualties during Tet—by some estimates, 40,000-50,000 killed across all areas of operation.

**5. "AI disruption... time available to adapt is compressing"** The chapter references research on technology adoption curves. The specific timeframes cited (Electricity ~50 years, Telephone ~50 years,

Television ~30 years, Internet ~15 years, Mobile ~10 years, Social media ~5 years, AI ~2 years) are illustrative estimates drawn from various technology adoption studies.

## CHAPTER 7: RED NOSES IN NORWAY

*Submarine Hunting and Product Development*

1. **"P-3C Orion... anti-submarine warfare"** The P-3 Orion was developed by Lockheed in the late 1950s based on the L-188 Electra commercial airliner. The P-3C variant, introduced in 1969, featured advanced sensors and data processing for submarine detection. The aircraft remained in U.S. Navy service until replaced by the P-8A Poseidon.

2. **"Sonobuoys"** Sonobuoys are expendable sensors dropped from aircraft into the ocean. They deploy hydrophone arrays to detect submarine acoustic signatures and transmit data back to the aircraft via radio. Types include passive (listening only), active (emitting sonar pulses), and specialized variants for bathymetric data and environmental conditions.

3. **"Akula-class... Russian nuclear submarine"** The Akula-class (Project 971 Shchuka-B) is a series of Soviet/Russian nuclear-powered attack submarines known for their quiet operation. The class entered service in 1986 and represented a significant advance in Soviet submarine technology.

4. **"thermal layer"** The thermocline is a layer in the ocean where temperature decreases rapidly with depth, creating an acoustic boundary that affects sound propagation. Submarines can use the thermocline to evade detection by operating above or below the depth where search sonobuoys are positioned.

**5. "Color launched with $41 million in funding... March 2011... shut down... October 2012"** Color Labs raised $41 million in venture funding before launching their proximity-based photo-sharing app. Despite the massive funding, the app failed to gain user adoption. Apple acquired the company's engineering team and patents in 2012 for approximately $7 million, a fraction of the original investment.

**6. "Pear Therapeutics... FDA approval... 2017... bankruptcy... April 2023"** Pear Therapeutics received FDA approval for reSET (for substance use disorders) in 2017, making it the first FDA-approved mobile app for treating disease. The company went public via SPAC in December 2021 at a $1.6 billion valuation. Despite FDA-approved products, Pear struggled to achieve reimbursement from health insurers and filed for Chapter 11 bankruptcy in April 2023. The company's assets were sold for approximately $6 million.

**7. "Netflix... completion rate"** Netflix's evolution from measuring "plays" to measuring completion rate reflected a broader shift in streaming analytics toward engagement metrics that predict customer satisfaction and retention, rather than simple usage counts.

## CHAPTER 8: THE COMMITTEE OF SAFETY

*Revolutionary War: Committees of Safety*

**1. "Committees of Safety became the shadow government"** Between 1774 and 1776, Committees of Safety assumed governmental functions as royal authority collapsed throughout the American colonies. These bodies organized militia, collected taxes, enforced trade boycotts, and maintained civil order. In North Carolina, county committees were coordinated through provincial congresses. See *The Committees of Safety of the American Revolution* by Agnes Hunt (1904, reprinted by various publishers).

**2. "The Pitt Association's language... pledged loyalty to King George III even as it promised to resist his policies"** The Pitt Association, like many similar documents from 1775, maintained a fiction of loyalty to the Crown while organizing armed resistance to royal policies. This rhetorical strategy reflected genuine uncertainty about whether reconciliation remained possible and served as legal cover if the revolutionary cause failed.

**3. "Commander's intent"** Commander's intent is a modern military doctrine concept that communicates the purpose and desired end state of an operation, allowing subordinates to adapt tactics while maintaining strategic alignment. The U.S. Army formally codified commander's intent doctrine in the 1980s, though the practice has historical antecedents.

**4. "Brown Bess surplus"** The "Brown Bess" was the British Land Pattern Musket used by British forces from 1722 to the mid-19th century. Colonial militia often used captured, purchased, or smuggled Brown Bess muskets, along with various hunting firearms and locally manufactured weapons.

**5. "Field return taken on October 12... 'Capt. George Moy, 58 effectives'"** Militia field returns recorded the strength and equipment status of companies. "Effectives" referred to men present and fit for duty, as opposed to those sick, absent, or lacking weapons.

**6. "Colonel Richard Caswell and Colonel Alexander Lillington"** Richard Caswell (1729-1789) later became the first governor of the independent State of North Carolina. Alexander Lillington

(1725-1786) commanded New Hanover County militia. Both played crucial roles in the Patriot victory at Moore's Creek Bridge.

## CHAPTER 9: WHEN THE CAPTAIN FALLS

*Civil War: The 57th Georgia Through Catastrophe*

1. **"Siege of Vicksburg... forty-seven days"** The Siege of Vicksburg lasted from May 18 to July 4, 1863 (technically 47 days). Confederate forces and civilians endured constant bombardment, dwindling rations, and epidemic disease before Pemberton's surrender.

2. **"Andersonville... Camp Sumter"** Camp Sumter (Andersonville Prison) was established in February 1864 in response to the breakdown of prisoner exchange arrangements. The 57th Georgia served guard duty at Andersonville in spring 1864 before being recalled to the front for the Atlanta Campaign. The camp's commander, Captain Henry Wirz, was the only Confederate official executed for war crimes after the war.

3. **"Union prisons like Elmira"** Elmira Prison in New York held Confederate prisoners of war from July 1864 to July 1865. Approximately 2,950 of 12,123 prisoners died, a mortality rate of 24%—comparable to Andersonville's 28%. Conditions at Northern prisons, while generally better than Southern facilities, were often brutal.

4. **"Battle of Kennesaw Mountain... June 27, 1864... frontal assault"** Sherman's decision to attack entrenched Confederate positions directly at Kennesaw Mountain departed from his typical flanking strategy. The assault's failure—3,000 Union casualties versus 800 Confederate—reinforced Sherman's preference for maneuver over direct attack.

**5. "Battle of Bentonville, North Carolina"** The Battle of Bentonville (March 19-21, 1865) was the last major Confederate offensive of the war. General Joseph E. Johnston concentrated scattered Confederate forces for a surprise attack on Sherman's advancing columns. After initial success, Confederate forces were pushed back. The 57th Georgia, by then consolidated into the 1st Georgia Consolidated Infantry, participated in this final engagement.

**6. "Tech layoffs of 2022-2023... Meta... Amazon... Google... Twitter"** The technology industry experienced widespread layoffs in late 2022 and 2023 as companies adjusted to post-pandemic market conditions and rising interest rates. Major layoffs included: Meta (11,000 in November 2022; 10,000 in March 2023); Amazon (27,000 across multiple rounds); Google (12,000 in January 2023); Twitter (approximately 3,700 immediately after acquisition, followed by additional cuts). See contemporary reporting in *The Wall Street Journal*, *TechCrunch*, and other technology publications.

## CHAPTER 10: THE LIAISON

*Vietnam War: Coalition Leadership and the Orphanage Project, 1967–1968*

**1. "The orphanage sat in a clearing three thousand meters southeast of Camp Bearcat's perimeter"** Camp Bearcat (officially Camp Martin Cox) was the base camp for the U.S. 9th Infantry Division, located in Long Thanh District, Bien Hoa Province, approximately 20 miles southeast of Saigon. The base was established in late 1966 and served as divisional headquarters until the 9th Division's departure in 1969. The camp's location placed it in a contested area where Viet Cong forces maintained significant presence in surrounding villages and hamlets.

**2. "It was February 1968, two weeks after the Tet Offensive had torn through Saigon"** The Tet Offensive began on January 30-31, 1968, during the Vietnamese lunar new year (Tế t Nguyên Đán). Approximately 70,000 North Vietnamese and Viet Cong troops launched coordinated attacks on more than 100 cities, towns, and military installations across South Vietnam. The offensive was a tactical defeat for communist forces—they suffered an estimated 30,000-50,000 killed—but a strategic victory in its impact on American public opinion. See James H. Willbanks, *The Tet Offensive: A Concise History* (Columbia University Press, 2007).

**3. "Captain James Moye III... civil affairs officer to the 2nd Battalion, 47th Infantry Regiment, 9th Infantry Division"** James Morris Moye III was born April 19, 1940, in Sanford, Florida. He served as an Army captain during the Vietnam War, deploying to Vietnam in late 1967 with the 9th Infantry Division. Civil affairs officers coordinated military operations with civilian populations, managed refugee assistance, and liaised with local Vietnamese officials—a role that required diplomatic skill and cultural sensitivity rather than direct combat command.

**4. "seventy-two children and a monk named Nguyễn Văn Siu"** Buddhist monasteries and temples throughout South Vietnam served as informal refugee centers during the war, particularly for children orphaned by combat. The role of Buddhist institutions in providing humanitarian assistance is documented in Robert J. Topmiller's *The Lotus Unleashed: The Buddhist Peace Movement in South Vietnam, 1964–1966* (University Press of Kentucky, 2002).

**5. "'We do not want your money. Money brings evil'"** The monk's skepticism reflected a broader pattern in Vietnamese society where American aid was frequently diverted by corruption.

The black market for American supplies was pervasive, and local officials often demanded bribes or kickbacks for access to aid. This corruption undermined American legitimacy and fueled Vietnamese resentment. The dynamics are examined in James M. Carter's *Inventing Vietnam: The United States and State Building, 1954–1968* (Cambridge University Press, 2008).

**6. "September 21, 1967, the first Thai combat forces deployed to South Vietnam... the 'Queen's Cobras'"** The Royal Thai Army Volunteer Regiment, nicknamed the "Queen's Cobras," arrived in Vietnam on September 21, 1967. The 2,200-soldier force was part of the "Free World Military Forces" program, through which the United States encouraged allied nations to contribute troops to the war effort. Thailand's participation was motivated by concerns about communist expansion in Southeast Asia and by substantial American financial support. See Robert M. Blackburn, *Mercenaries and Lyndon Johnson's "More Flags": The Hiring of Korean, Filipino, and Thai Soldiers in the Vietnam War* (McFarland, 1994).

**7. "The Thais... set up alongside the U.S. 9th Infantry Division"** Thai forces established their headquarters at Camp Bearcat alongside the American 9th Infantry Division. The arrangement required constant coordination between forces with different chains of command, languages, and military traditions. Thai commanders reported to Bangkok, not to MACV (Military Assistance Command, Vietnam), creating the coalition coordination challenges described in this chapter.

**8. "MACV (Military Assistance Command, Vietnam)"** MACV was the unified command structure for American military operations in Vietnam from 1962 to 1973. Headquartered in Saigon, MACV coordinated U.S. forces, advised the South Vietnamese military, and liaised with allied "Free World" forces. General

William Westmoreland commanded MACV from 1964 to 1968, followed by General Creighton Abrams. The command's organizational history is documented in Graham A. Cosmas, *MACV: The Joint Command in the Years of Escalation, 1962–1967* (U.S. Army Center of Military History, 2006).

**9. "On the night of January 31, 1968, Viet Cong forces launched simultaneous attacks across South Vietnam"** The Tet Offensive's coordinated timing—designed to achieve maximum shock—represented one of the largest military operations of the war. In III Corps (the military region surrounding Saigon), major targets included Tan Son Nhut Air Base, the U.S. Embassy, the Presidential Palace, and military installations throughout Saigon and surrounding provinces. The National Archives maintains extensive documentation at https://www.archives.gov/research/military/vietnam-war/tet-offensive.

**10. "In III Corps, they hit Saigon, the Long Binh logistics complex, and Bien Hoa Airbase"** Long Binh Post was the largest U.S. Army installation in Vietnam, serving as the logistics and administrative hub for II Field Force Vietnam. The base housed ammunition depots, supply warehouses, and the headquarters of multiple commands. During Tet, Viet Cong sappers penetrated the perimeter and detonated ammunition bunkers, creating explosions visible for miles. The base defense and counterattack are documented in Erik Villard, *The 1968 Tet Offensive Battles of Quang Tri City and Hue* (U.S. Army Center of Military History, 2008).

**11. "The battalion raced north in their M113 armored personnel carriers"** The M113 armored personnel carrier was the standard infantry transport vehicle in Vietnam. The 2/47th Infantry was a mechanized battalion, meaning its soldiers fought mounted in M113s rather than on foot. The aluminum-armored vehicles provided protection against small arms fire but were vulnerable to mines and rocket-propelled grenades. In urban combat during Tet, the vehicles' armor made them valuable for clearing streets, though they were poorly suited for the close-quarters fighting in Saigon's dense neighborhoods.

**12. "redirected into Saigon itself to help dislodge enemy forces from the Cholon district"** Cholon, Saigon's Chinatown district, saw some of the heaviest urban fighting of the Tet Offensive. Viet Cong forces occupied buildings, established defensive positions, and fought room-to-room against American and South Vietnamese forces. The 2/47th Infantry, trained for mechanized warfare in open terrain, found themselves in unfamiliar urban combat. Keith William Nolan's *House to House: Playing the Enemy's Game in Saigon, May 1968* (Zenith Press, 2006) covers the urban fighting during both the January offensive and "Mini-Tet" in May.

**13. "By mid-March, MACV created the Royal Thai Army Volunteer Force – Special Liaison Section (RTAVF-SLS)"** The Special Liaison Section consisted of approximately 18 U.S. personnel assigned to coordinate American support for Thai operations. SLS personnel were embedded at Thai firebases and accompanied Thai units on operations. Their responsibilities included coordinating artillery fire support, close air support, medical evacuation, intelligence sharing, and logistics. The creation of dedicated liaison units reflected lessons learned from the coordination challenges of the initial Thai deployment.

**14. "American F-4 Phantoms would be overhead"** The McDonnell Douglas F-4 Phantom II was the primary American fighter-bomber in Vietnam, capable of delivering bombs, napalm, and other ordnance in close air support of ground forces. Coordinating close air support for allied forces required precise communication of target locations and friendly positions—the core function of the SLS.

**15. "a seventy-two-foot well producing clean water"** Access to clean water was a critical need in rural Vietnam, where contaminated surface water contributed to disease. American military well-drilling teams operated throughout South Vietnam as part of civic action programs designed to win "hearts and minds." The well described in the chapter provided the orphanage with a reliable source of uncontaminated water, eliminating dependence on the nearby creek.

**16. "'These are still people. That's what we're here for'"** *Stars and Stripes* is the independent daily newspaper serving the U.S. Armed Forces, published continuously since World War II. The newspaper's Vietnam coverage included features on soldiers' humanitarian efforts alongside combat reporting. Captain Moye's statement reflects the tension between the war's destructive mission and individual soldiers' impulse toward humanitarian action—a theme explored in Christian G. Appy's *Working-Class War: American Combat Soldiers and Vietnam* (University of North Carolina Press, 1993).

**17. "The Thais would eventually withdraw in 1971"** Thai forces withdrew from Vietnam in 1971 as part of the broader "Vietnamization" policy under President Nixon, which aimed to transfer combat responsibilities to South Vietnamese forces while reducing American and allied troop levels. At peak strength, approximately 11,000 Thai troops served in Vietnam, including the Black Panther Division that arrived in 1968. Thai forces suffered approximately 350 killed in action during their deployment. See Stanley Robert Larsen and James Lawton Collins Jr., *Allied Participation in Vietnam* (Department of the Army, 1975).

**18. "The orphanage... survived the war and became a permanent institution"** Many Buddhist institutions that served as informal refugee centers during the war continued their humanitarian missions after 1975. While specific documentation of this particular orphanage's post-war history is limited, the pattern of Buddhist institutions providing social services persisted through the communist takeover and continues today.

**19. "In 2003, a small biotechnology company called Genentech was working on Avastin"** Genentech, founded in 1976, pioneered the biotechnology industry and developed numerous breakthrough drugs including human insulin, human growth hormone, and cancer treatments. By 2003, Genentech was developing bevacizumab (brand name Avastin), a monoclonal antibody that inhibits tumor blood vessel growth (angiogenesis). The drug represented a new approach to cancer treatment—targeting the tumor's blood supply rather than the cancer cells directly.

**20. "Avastin... one of the best-selling oncology treatments in history"** The FDA approved Avastin for metastatic colorectal cancer in February 2004. Subsequent approvals expanded its use to lung cancer, kidney cancer, brain tumors, and other malignancies. At its peak, Avastin generated over $7 billion in annual global sales, making it one of the highest-revenue drugs in pharmaceutical history. The drug extended survival for many cancer patients, though later studies prompted FDA to revoke approval for breast cancer indication in 2011.

**21. "academic researchers... clinical trial sites... contract manufacturers... regulatory agencies... hospitals... patients"** The modern drug development ecosystem involves hundreds of organizations across multiple sectors. Academic researchers (primarily funded by NIH and other government grants) discover biological mechanisms and potential drug targets. Biotechnology and pharmaceutical companies develop drug candidates. Contract research organizations (CROs) manage clinical trials. Contract manufacturing organizations (CMOs) produce drugs at scale. The FDA evaluates safety and efficacy. Hospitals and physicians administer treatments. This fragmented ecosystem requires coordination rather than command.

**22. "Genentech functioned as a coordinator, not a commander—precisely like the SLS at Camp Bearcat"** Genentech's organizational model emphasized collaboration with academic researchers, treating them as partners rather than contractors. The company maintained close relationships with leading cancer researchers at major academic medical centers, funding research while sharing credit and intellectual property. This approach attracted top scientific talent and accelerated drug development. The model is discussed in Gordon Binder and Philip Bashe, *Science Lessons: What the Business of Biotech Taught Me About Management* (Harvard Business Press, 2008).

**23. "They organized teams around diseases, not departments"** Genentech's cross-functional team structure—organizing around therapeutic areas (oncology, immunology, etc.) rather than traditional corporate functions—became a model for the biotechnology industry. These "disease teams" included researchers, clinicians, regulatory specialists, manufacturing experts, and commercial personnel working together from early development through product launch. The approach reduced handoff delays and ensured that commercial and regulatory considerations informed scientific decisions early in development.

**24. "They also engaged with the FDA early and often"** Genentech pioneered the practice of frequent pre-submission meetings with

FDA, treating the regulatory process as a collaborative dialogue rather than an adversarial review. This approach—now standard industry practice—allowed companies to design clinical trials that would meet FDA requirements, reducing the risk of rejection after years of expensive development. The FDA's guidance on pre-IND meetings and other collaborative processes is available at https://www.fda.gov/drugs/investigational-new-drug-ind-application/meetings-ind-sponsors.

**25. "Avastin received FDA approval in 2004 and became a multi-billion-dollar drug"** The FDA approved Avastin on February 26, 2004, for first-line treatment of metastatic colorectal cancer in combination with chemotherapy. The approval came approximately six months after Genentech submitted its Biologics License Application—a relatively fast review reflecting the drug's strong clinical data and the FDA's priority review designation for serious conditions. The approval history and prescribing information are available at https://www.accessdata.fda.gov/drugsatfda_docs/label/2004/125085lbl.pdf.

## CHAPTER 11: THE EMPOWERED CREW

*Iraq War: The Baghdad SAR Mission, April 6, 2003*

**1. "The P-3C Orion banked left over the black water south of Crete"** The P-3C Orion is a four-engine turboprop maritime patrol aircraft developed by Lockheed (now Lockheed Martin) for the United States Navy. Originally designed in the 1960s for anti-submarine warfare during the Cold War, the aircraft was adapted for overland Intelligence, Surveillance, and Reconnaissance (ISR) missions during Operation Iraqi Freedom. P-3 squadrons operated from Naval Air Station Sigonella in Sicily and Souda Bay Naval Air Station in Crete during the 2003 invasion. The Naval History and Heritage Command maintains extensive documentation on naval aviation operations: https://www.history.navy.mil/

**2. "It was April 6, 2003"** By April 6, 2003, Operation Iraqi Freedom was in its third week. Coalition ground forces had advanced rapidly toward Baghdad, with the 3rd Infantry Division reaching the city's outskirts. The famous "Thunder Run" armored assault into Baghdad would occur the following day, April 7. The operational tempo placed enormous demands on ISR assets, which were critical for identifying enemy positions, tracking vehicle movements, and supporting search-and-rescue operations. See Michael R. Gordon and Bernard E. Trainor, *Cobra II: The Inside Story of the Invasion and Occupation of Iraq* (Vintage, 2007).

**3. "Combat Air Crew Four—CAC-4"** Navy patrol squadrons organize their personnel into numbered Combat Air Crews, each consisting of approximately 11-12 personnel who train and fly together as a unit. This crew concept builds the cohesion and mutual trust essential for effective operations. A typical P-3 crew includes three pilots (aircraft commander, co-pilot, and third pilot), two Naval Flight Officers (tactical coordinator and navigator/communicator), three sensor operators, two flight engineers, and one in-flight technician.

**4. "providing Intelligence, Surveillance, and Reconnaissance (ISR) support for ground forces"** ISR missions involve collecting information about enemy forces, terrain, and activities to support military decision-making. During Operation Iraqi Freedom, ISR platforms tracked Iraqi military movements, identified targets, assessed battle damage, and supported special operations. The P-3's sensors—originally designed to detect submarines—were adapted to track vehicles, monitor communications, and provide real-time imagery to ground commanders.

**5. "The P-3, a Cold War relic designed for anti-submarine warfare, had been hastily retrofitted"** The P-3 Orion entered Navy service in 1962, designed primarily to hunt Soviet submarines using sonobuoys, magnetic anomaly detection, and torpedoes. By 2003, the aircraft was being replaced by the P-8 Poseidon in the anti-submarine role. However, its long endurance (10+ hours), sensor capacity, and availability made it valuable for overland ISR when purpose-built platforms were in short supply. The adaptation exemplifies military improvisation under operational pressure.

**6. "tactical coordinator for CAC-4"** The Tactical Coordinator (TACCO) is a Naval Flight Officer who serves as the mission commander for maritime patrol aircraft. While the aircraft commander (a pilot) is responsible for safe operation of the aircraft, the TACCO directs the tactical mission: determining search patterns, prioritizing sensor employment, coordinating with external agencies, and making real-time decisions about mission execution. This division of authority—pilot flies the airplane, TACCO runs the mission—exemplifies the distributed leadership model described throughout this chapter.

**7. "the aircraft commander, a lieutenant commander we called 'G-Money'"** Military aviators traditionally use callsigns—nicknames used over radio and in informal settings. Callsigns are typically assigned by squadron mates and often reference personality traits, memorable incidents, or physical characteristics. The practice builds unit cohesion and provides operational security by avoiding use of real names over radio communications.

**8. "Three pilots rotated flying duties. Two naval flight officers managed sensors and navigation. Three sensor operators ran radar and cameras. Two flight engineers monitored engines and fuel. One in-flight technician watched for threats"** The P-3C crew composition reflects the aircraft's complex mission requirements. Flight engineers (enlisted Aviation Machinist's Mates) manage the four Allison T56 turboprop engines, fuel systems, and auxiliary power systems. Sensor operators (enlisted Aviation Warfare Systems Operators) run the aircraft's radar, cameras, electronic surveillance equipment, and acoustic sensors. In-flight technicians monitor defensive systems and maintain situational awareness through visual observation. This specialization creates interdependence—no single crew member can perform all functions, requiring trust and coordination.

**9. "'Madman Five-Seven, this is Magic'"** "Madman" was the callsign for P-3 aircraft from Patrol Squadron Five (VP-5), nicknamed the "Mad Foxes." "Magic" was a common callsign for AWACS controllers. The AWACS—Boeing E-3 Sentry aircraft—served as airborne command and control platforms, coordinating multiple aircraft and providing radar coverage over the battlefield.

**10. "A coalition helicopter is down near Baghdad. Search and rescue is inbound"** Combat Search and Rescue (CSAR) operations during Operation Iraqi Freedom were primarily conducted by Air Force HH-60 Pave Hawk helicopters and pararescue personnel (PJs). When an aircraft was downed, the priority was locating survivors before enemy forces could capture them. ISR platforms like the P-3 could provide critical situational awareness—identifying the crash site, tracking enemy movements, and guiding rescue helicopters to safe approach routes. The interservice coordination described in this chapter was facilitated by AWACS aircraft serving as airborne command posts.

**11. "The P-3 had no defensive weapons, no countermeasures beyond chaff and flares"** Chaff consists of small metallic strips ejected from aircraft to create false radar returns, potentially confusing radar-guided missiles. Flares are high-temperature decoys designed to attract heat-seeking missiles away from the aircraft's engines. These passive countermeasures were the P-3's only defenses against surface-to-air threats. The aircraft carried no guns, missiles, or other offensive weapons when configured for ISR missions.

**12. "Baghdad was defended by anti-aircraft artillery and missiles"** Iraqi air defenses in April 2003 included radar-guided surface-to-air missiles (SA-2, SA-3, SA-6), man-portable infrared missiles (SA-7, SA-14, SA-16), and extensive anti-aircraft artillery (AAA). While coalition Suppression of Enemy Air Defenses (SEAD) operations had degraded Iraqi capabilities, significant threats remained, particularly from mobile systems and man-portable missiles that could be concealed and rapidly deployed.

**13. "We didn't join the Navy to fly safe missions. We joined because we took an oath"** The oath of office for commissioned officers in the United States Armed Forces reads: "I do solemnly swear that I will support and defend the Constitution of the United States against all enemies, foreign and domestic; that I will bear true faith and allegiance to the same; that I take this obligation freely, without any mental reservation or purpose of evasion; and that I will well and faithfully discharge the duties of the office on which I am about to enter. So help me God." The oath's emphasis on the Constitution rather than any individual leader reflects the Founders' intent to subordinate military power to civilian authority.

**14. "Wreckage visible, looks like a Blackhawk"** The UH-60 Black Hawk is the U.S. Army's primary utility helicopter, used for troop transport, medical evacuation, and other missions. Several helicopters were lost during Operation Iraqi Freedom, including to enemy fire and accidents. The specific incident described in this chapter—a helicopter down southwest of Baghdad requiring ISR overwatch—illustrates the dynamic nature of combat operations and the improvised tasking that characterized the early weeks of the invasion.

**15. "don't enter Iranian airspace, don't drop below 15,000 feet"** These operational constraints reflect the complex airspace management of the Iraq theater. Iranian airspace was strictly off-limits to avoid international incidents. The 15,000-foot minimum altitude placed the aircraft above the effective range of most man-portable air defense systems (MANPADS) while still allowing sensors to acquire useful imagery. These constraints defined the boundaries within which the crew had autonomy to execute the mission.

**16. "Bingo fuel: the minimum required to return to base"** "Bingo fuel" is the calculated fuel quantity at which an aircraft must depart the operating area to safely reach its recovery base, including required reserves. The calculation accounts for distance, weather, potential diversions, and regulatory fuel reserves. Continuing operations past bingo fuel commits the aircraft to diverting to an alternate field or risking fuel exhaustion—a decision with potentially fatal consequences.

**17. "We can shut down one engine. Reduces fuel burn by about 25 percent"** The P-3C can safely operate on three of its four Allison T56-A-14 turboprop engines, though with degraded performance. Each engine produces approximately 4,600 shaft horsepower. Shutting down one engine reduces fuel consumption significantly but also reduces available thrust for climbing, accelerating, or maneuvering. In a threat environment, this trade-off accepts increased vulnerability to extend mission duration.

**18. "We diverted to Saudi Arabia, landed at a small coalition airfield, refueled"** Coalition forces operated from multiple airfields throughout the Persian Gulf region during Operation Iraqi Freedom, including bases in Saudi Arabia, Kuwait, Qatar, Bahrain, and the United Arab Emirates. Diverting to an unplanned recovery field required coordination with air traffic control and host nation authorities, but was a routine contingency for which crews trained. The ability to adapt to changing circumstances—landing where fuel was available rather than insisting on returning to the planned base—exemplifies operational flexibility.

**19. "In 1999, Tony Hsieh invested in Zappos"** Tony Hsieh (1973–2020) initially invested in Zappos in 1999 and became CEO in 2000. A Harvard computer science graduate, Hsieh had previously co-founded LinkExchange, which Microsoft acquired for $265 million in 1998. Under his leadership, Zappos grew from a struggling startup to a billion-dollar company known for exceptional customer service. Hsieh documented his management philosophy in *Delivering Happiness: A Path to Profits, Passion, and Purpose* (Business Plus, 2010). Hsieh died in a house fire in November 2020.

**20. "selling shoes online was considered impossible"** In the late 1990s, conventional retail wisdom held that shoes could not be sold online because customers needed to try them on for fit. Zappos overcame this objection through free shipping, free returns, and a 365-day return policy—effectively letting customers try shoes at home with no risk. This customer-centric approach required trusting that most customers would not abuse the generous policies, a bet that proved correct.

**21. "hired customer service reps in Las Vegas"** Zappos relocated its headquarters and call center from San Francisco to Las Vegas in 2004, attracted by lower costs of living and a large pool of service-industry workers accustomed to customer-facing roles. The Las Vegas location became central to company culture. Hsieh later invested heavily in downtown Las Vegas revitalization through his Downtown Project, attempting to apply Zappos principles to urban development.

**22. "complete discretion to do whatever they think will create a remarkable experience"** Zappos customer service representatives operated without scripts and were not measured on call duration—metrics that dominated most call centers. Instead, they were evaluated on customer satisfaction and empowered to make decisions that would create stories worth telling. This approach inverted the traditional cost-center mentality of customer service, treating it instead as a marketing and brand-building investment.

**23. "A rep sent flowers to a customer whose mother had just died"** Zappos customer service stories became legendary examples of empowered decision-making. The flowers story, the ten-hour phone call, the wedding shipping upgrade, and the competitor referral all represented individual employees making judgment calls that cost money short-term but built brand loyalty long-term. These anecdotes circulated widely in business media and became case studies in customer experience design. See Joseph A. Michelli, *The Zappos Experience: 5 Principles to Inspire, Engage, and WOW* (McGraw-Hill, 2011).

**24. "Zappos grew from $1.6 million in sales in 2000 to $1 billion in 2008"** Zappos's growth trajectory demonstrated that customer-centric culture could drive business results. The company reached $1 billion in gross merchandise sales by 2008, making it one of the fastest-growing e-commerce companies of its era. Growth came primarily through word-of-mouth and repeat customers rather than heavy advertising spending—validating Hsieh's thesis that remarkable experiences generate organic marketing.

**25. "Amazon acquired them in 2009 for $1.2 billion"** Amazon acquired Zappos in November 2009 for approximately $1.2 billion in stock. Unusually for Amazon acquisitions, the company allowed Zappos to continue operating independently, preserving its distinctive culture. Jeff Bezos stated that Amazon wanted to learn from Zappos's customer service approach rather than impose Amazon's practices. The acquisition validated Hsieh's thesis that culture and customer experience could create substantial enterprise value. See Brad Stone, *The Everything Store: Jeff Bezos and the Age of Amazon* (Little, Brown, 2013).

**26. "New hires spent four weeks learning company culture, product knowledge, and customer service philosophy"** Zappos's extended onboarding process was unusual in an industry known for minimal training and high turnover. The four-week program included classroom instruction, shadowing experienced representatives, and immersion in company values. This investment in training preceded the grant of autonomy—competence first, then empowerment. The approach reflected Hsieh's understanding that empowerment without capability produces chaos rather than results.

**27. "At the end of the first week, Zappos offered every new hire $2,000 to quit"** The "pay to quit" offer (later increased to $3,000–$4,000) served multiple purposes: it selected for commitment, it gave employees agency in their choice to stay, and it ensured that those who remained were actively engaged rather than passively employed. Approximately 10 percent of new hires took the offer. Amazon later adopted a similar practice for warehouse workers, offering up to $5,000 to resign, calling it "Pay to Quit."

**28. "The CEO is the Chief Repetition Officer"** The concept that leaders must communicate purpose relentlessly appears in multiple leadership frameworks. Patrick Lencioni discusses the need for "cascading communication" in *The Advantage* (Jossey-Bass, 2012). Simon Sinek's "Start With Why" framework similarly emphasizes that purpose must be constantly reinforced. The military concept of "commander's intent"—stating the purpose clearly enough that subordinates can make independent decisions aligned with that purpose—reflects the same principle.

**29. "After every mission, the P-3 crew conducted a debrief"** After-action reviews (AARs) originated in the U.S. Army and have been adopted across military services and civilian organizations. The AAR format—What was supposed to happen? What actually happened? Why was there a difference? What will we do differently?—focuses on learning rather than blame. The U.S. Army's Center for Army Lessons Learned maintains extensive resources on AAR methodology. See also Marilyn Darling, Charles Parry, and Joseph Moore, "Learning in the Thick of It," *Harvard Business Review* (July–August 2005).

**30. "Organizations that master after-action reviews build institutional learning that compounds over time"** The concept of organizational learning—institutions improving their performance through systematic reflection and adaptation—is explored in Peter Senge's *The Fifth Discipline: The Art and Practice of the Learning Organization* (Doubleday, 1990). The military's emphasis on "lessons learned" systems reflects this principle, though implementation varies widely across organizations.

# *Sources for Further Reading*


## REVOLUTIONARY WAR:

Buchanan, John. *The Road to Guilford Courthouse: The American Revolution in the Carolinas* (Wiley, 1997)

Morrill, Dan L. *The Southern Campaigns of the American Revolution* (1992)

Rankin, Hugh F. "The Moore's Creek Bridge Campaign, 1776," *North Carolina Historical Review* 30 (January 1953)

Carbone, Gerald M. *Nathanael Greene: A Biography of the American Revolution* (2008)

## CIVIL WAR:

Ballard, Michael B. *Vicksburg: The Campaign That Opened the Mississippi* (University of North Carolina Press, 2004)

Marvel, William. *Andersonville: The Last Depot* (University of North Carolina Press, 1994)

McMurry, Richard M. *Atlanta 1864: Last Chance for the Confederacy* (University of Nebraska Press, 2000)

## VIETNAM WAR:

Larsen, Stanley Robert, and James Lawton Collins Jr. *Allied Participation in Vietnam* (Department of the Army, 1975)

Willbanks, James H. *The Tet Offensive: A Concise History* (Columbia University Press, 2007)

Stanton, Shelby L. *The Rise and Fall of an American Army: U.S. Ground Forces in Vietnam, 1965-1973* (Dell, 1985)

Schreadley, R.L. *The Riverine Force* (1969)

Blackburn, Robert M. *Mercenaries and Lyndon Johnson's "More Flags": The Hiring of Korean, Filipino, and Thai Soldiers in the Vietnam War*. McFarland, 1994.

Cosmas, Graham A. *MACV: The Joint Command in the Years of Escalation, 1962–1967*. U.S. Army Center of Military History, 2006.

Larsen, Stanley Robert, and James Lawton Collins Jr. *Allied Participation in Vietnam*. Department of the Army, 1975.

Nolan, Keith William. *House to House: Playing the Enemy's Game in Saigon, May 1968*. Zenith Press, 2006.

Willbanks, James H. *The Tet Offensive: A Concise History*. Columbia University Press, 2007.

## IRAQ WAR:

Gordon, Michael R., and Bernard E. Trainor. *Cobra II: The Inside Story of the Invasion and Occupation of Iraq* (Vintage, 2007)

Ricks, Thomas E. *Fiasco: The American Military Adventure in Iraq* (Penguin Press, 2006)

## BUSINESS AND LEADERSHIP:

Horowitz, Ben. *The Hard Thing About Hard Things* (Harper Business, 2014)

Hsieh, Tony. *Delivering Happiness* (Business Plus, 2010)

Gerstner, Louis V., Jr. *Who Says Elephants Can't Dance?* (Harper Business, 2002)

McChrystal, Stanley. *Team of Teams: New Rules of Engagement for a Complex World* (Portfolio, 2015)

Sinek, Simon. *Start With Why: How Great Leaders Inspire Everyone to Take Action*. Portfolio, 2009.

Lencioni, Patrick. *The Advantage: Why Organizational Health Trumps Everything Else in Business*. Jossey-Bass, 2012.

Senge, Peter. *The Fifth Discipline: The Art and Practice of the Learning Organization*. Doubleday, 1990.

U.S. Army Center for Army Lessons Learned: https://usacac.army.mil/organizations/mccoe/call

## PRIMARY SOURCES:

Saunders, William L., ed. *The Colonial Records of North Carolina*, 10 volumes (Raleigh: State of North Carolina, 1886-1890). Available digitally through Documenting the American South, University of North Carolina.

Clark, Walter, ed. *The State Records of North Carolina*, 16 volumes (Winston and Goldsboro: State of North Carolina, 1895-1906)

National Archives, Vietnam War records: https://www.archives.gov/research/military/vietnam-war

FDA Drug Approval Database: https://www.accessdata.fda.gov/scripts/cder/daf/

Vietnam Center and Sam Johnson Vietnam Archive, Texas Tech University: https://www.vietnam.ttu.edu

## NATIONAL PARK SERVICE SITES:

Moores Creek National Battlefield: https://www.nps.gov/mocr/

Guilford Courthouse National Military Park: https://www.nps.gov/guco/

Vicksburg National Military Park: https://www.nps.gov/vick/

Andersonville National Historic Site: https://www.nps.gov/ande/

Kennesaw Mountain National Battlefield Park: https://www.nps.gov/kemo/

## DIGITAL ARCHIVES:

Fold3 (military records): https://www.fold3.com

National Archives: https://www.archives.gov

Vietnam Center and Sam Johnson Vietnam Archive, Texas Tech University: https://www.vietnam.ttu.edu

Documenting the American South: https://docsouth.unc.edu

## BIOTECHNOLOGY AND DRUG DEVELOPMENT:

Binder, Gordon, and Philip Bashe. *Science Lessons: What the Business of Biotech Taught Me About Management.* Harvard Business Press, 2008.

Pisano, Gary P. *Science Business: The Promise, the Reality, and the Future of Biotech.* Harvard Business School Press, 2006.

Robbins-Roth, Cynthia. *From Alchemy to IPO: The Business of Biotechnology.* Basic Books, 2001.

## NAVAL AVIATION:

Naval History and Heritage Command: https://www.history.navy.mil/

Gillcrist, Paul T. *Feet Wet: Reflections of a Carrier Pilot.* Pocket Books, 1990.

www.ingramcontent.com/pod-product-compliance
Lightning Source LLC
LaVergne TN
LVHW041108080826
845145LV00007B/1732